MW00712977

MODERN SAFETY AND RESOURCE CONTROL MANAGEMENT

MODERN SAFETY AND RESOURCE CONTROL MANAGEMENT

THOMAS D. SCHNEID
Department of Loss Prevention and Safety
Eastern Kentucky University

A Wiley-Interscience Publication
JOHN WILEY & SONS, INC.
New York • Chichester • Weinheim • Brisbane • Singapore • Toronto

This book is printed on acid-free paper. ♾

Published simultaneously in Canada.

For ordering and customer service, call 1-800-CALL-WILEY.

Library of Congress Cataloging-in-Publication Data

Schneid, Thomas D.
Modern safety and resource control management / Thomas D. Schneid.
p. cm.
"Wiley-Interscience publication."
Includes index.
ISBN 0-471-33118-X (alk. paper)
1. Industrial safety—Management. I. Title.
T55.S328 2000
658.3′82—dc21 99-29163
CIP

Printed in the United States of America.

10 9 8 7 6 5 4 3 2 1

CONTENTS

PREFACE

> There ain't no constants in business because people keep changing.
>
> —Solomon Dutka

> We must always change, renew, rejuvenate ourselves; otherwise we harden.
>
> —Johann Wolfgang von Goethe

The safety function has changed significantly since the inception of the Occupational Safety and Health Act in 1970 and the per se birth of the modern safety and health professional. The Occupational Safety and Health Act is relatively new at the age of 29; however, the functions of the safety and health professional have evolved and changed significantly during this period of time.

In general, the functions of the safety and health professional in the 1970s were primarily compliance driven with the activities and responsibilities focused on acquiring and maintaining the work environment within the boundaries specified in the Occupational Safety and Health standards. Given the relatively broad scope of the Occupational Safety and Health Act, newness of this type of legislation, impact on the management team, retrofitting of equipment, and time required within the learning curve, often required the full-time efforts of the safety and health professional. In essence, the safety and health professional was a single-function manager with all efforts focused on the essential work of achieving and maintaining compliance with this new law.

In the 1980s, the functions and job responsibilities of many safety and health professionals were broadened to include peripheral job functions, which often interacted with safety and health such as workers' compensation, security, and environmental compliance. In many organizations, compliance with the Occupational Safety and Health Act had been achieved through their years of effort and the organizations moved from a reactive to a proactive mode in the safety and health area. However, peripheral functions, especially workers' compensation, began to emerge as a substantial cost item for many organizations. Given the logic that it was the safety and health function that was supposed to prevent the accident and if the accident was not prevented, the injury, and thus the cost, was incurred, the safety and health professional should also assume responsibility for the reactive portion of the process. Although this marriage appeared symbiotic, the actual management was often like oil and water. Many safety and health professionals found that the majority of their time was expended on the reactive management of workers' compensation claims with a minimal amount of time being allotted to the proactive management of safety and health.

With the emergence of the environmental area, many safety and health professionals were saddled with the responsibility of achieving and maintaining the compliance activities and efforts in this evolving area. As one safety and health professional stated, "My management team looked at the environmental regulations. These regulations looked like the OSHA standards so they gave the responsibility to me." Although some of the environmental standards dovetail with the OSHA standards, many other environmental laws require substantial time and effort to achieve and maintain compliance. However, environmental compliance is another cost category for many organizations which required appropriate management and specific expertise, and it was often considered "cost effective" to provide this responsibility to the safety and health professional, rather than incur the expense of hiring another professional. In essence, the management team provided the safety and health professional with "another hat" (more responsibilities) to perform within the organization. With this and additional responsibilities, the time, and often the nerves, of the safety and health professional were stretched to the maximum.

During the 1990s, the safety and health profession continued to expand and develop in many new areas. The driving force for safety and health for many organizations evolved from a compliance-driven entity to a dollar-driven function. Many companies possessed little or no fear of OSHA because they had achieved and were maintaining the necessary level of compliance. However, the dollars being expended through the workers' compensation and insurance corridor had emerged as a major cost item for many companies. The safety and health function began to be driven to reduce these costs areas through proactive management of these correlating risk areas.

For many organizations, basic compliance with the Occupational Safety and Health Act was a "given" with the emphasis being placed on expanding and refining the activities that impacted the "bottom line" not only in safety and health but also in the correlating areas. Additionally, the function and responsibilities of many safety and health professionals were expended via the regulatory route to encompass new and emerging risk issues such as workplace violence, tuberculosis, and ergonomics, which do not "fit the mold" of a traditional safety and health program. Given the changing emphasis and regulatory expansion, the boundaries of responsibility for many safety and health professionals metamorphosized again.

We are now at the end of the 1990s looking forward to the twenty-first century. What is the safety and health profession going to look like in the year 2020? Think about how our world has evolved during the past 29 years.

- How many safety and health professionals used a computer in 1970? Today most professionals own at least one computer, and the computer has become an essential tool in performing our jobs.
- What was television in 1970? Today, we have virtually instantaneous news through such services as CNN as well as the internet.
- How did safety and health professionals communicate in 1970? In addition to the basic land telephone lines, we now have cellular phones, beepers, teleconferencing, E-mail, and numerous other electronic services.

- Are the hazards different today from 1970? Although the basic workplace hazards can be the same as in 1970, safety and health professionals are now requested to address new areas such as cumulative trauma illnesses, airborne hazards (such as tuberculosis), bloodborne pathogens (such as HBV), laser hazards, new chemical hazards, radiation hazards, and a myriad of new areas.
- Are employees different today than in 1970? The employee of today is more informed and possesses access to substantially more information than the employee in 1970. Employees today are substantially more mobile, less likely to spend their entire career with one organization, and possess expectations for items such as day care, wellness centers, and car pools, which were unheard of in 1970.
- Are employers different today than in 1970? In many circumstances, the employers hired a safety and health professional in 1970 simply to keep OSHA off their back. Today, employers expect a safety and health professional to protect a wide range of resources in a cost effective manner and to reduce or eliminate any risk that may hinder their operations. Additionally employers today are larger because of mergers and acquisitions and function on a "tightened clock," whereby long-term means tomorrow.

Where is the safety and health professional going to be, and what are his or her duties going to be in the future? The label may be the same, but the functions will be as different as day and night from 1970. As espoused in this text, the safety and health professional of the past will evolve into the safety and resource control professional of the future who will be expected not only to manage the basic compliance functions in the areas of environment and safety but also to be responsible for the proactive management of a broad spectrum of other risks and assets in the workplace, ranging from human assets through efficacy risks, in order to impact the bottom line.

WHY SAFETY AND RESOURCE CONTROL MANAGEMENT?

What is the best title to place on the actual functions and activities that are performed by what is commonly referred to as a safety department? The one consistency our research identified is the lack of consistency in the terminology used in the American workplace.

Traditionally, the safety function was primarily focused on OSHA compliance and activities in the injury and illness prevention area and thus the titles commonly utilized included safety manager, safety engineer, and safety coordinator. As responsibilities in periferial areas expanded, bifurcated titles such as safety/environmental manager and safety and workers' compensation administrator became commonplace. As additional insurance responsibilities were added to the function, titles such as safety and risk manager and safety and insurance administrator were often utilized. With the emergence of ISO, the title of the function evolved to include safety and quality director and safety and ISO manager. With the responsibilities for security and related areas, the terminology safety and loss prevention and loss prevention specialists became vogue in many industries. And most recently we have see the development of another hybrid, namely safety and personnel manager or safety and human resources director.

To this end, we attempted to identify as many of the other "hats" that have been placed on the traditional industrial safety function in many companies. These additional duties in general categories include the following:

- Inspection
- Fire protection/suppression
- Security
- Environmental compliance
- Personnel and human resources
- Workers' compensation
- Insurance (e.g., health, property)
- Quality assurance
- Hazardous materials
- Fleet safety
- Industrial hygiene
- ISO
- Labor liaison
- Governmental liaison
- Other governmental requirements (such as HCCP, FDA, etc.)
- Risk management

Given these various and varied responsibilities, a search was conducted to identify the label that best identified the actual job function of this new hybrid entity. While not rejecting the traditional labels of safety manager and safety director as well as the newer terminology such as loss prevention, the term *safety and resource control management* was identified as a better description of the job function today.

Will the label safety and resource control management change the actual job functions in any given company? The answer is probably "no." However, this "label" does provide more clarity and better describes the numerous and varied job functions that have been or can be housed within the traditional safety function and provides a better explanation as to what modern safety and health professionals actually do in their jobs. The positive label of safety and resource control management encompasses not only the traditional safety duties but also the myriad of control activities for various human, equipment, efficacy, and related resources within the organization. Safety and resource control management manages and controls the various resources and potential risks related to each various organizational resource proactively in a cost-effective and efficient manner designed to achieve optimal results.

ACKNOWLEDGMENTS

Special thanks to my wife, Jani, and my children, Shelby, Madison, and Kasi, for their indulgence, support, and time during the writing of this text.

My thanks to my parents for their guidance and leadership, which has permitted me to acquire the education and skills to be able to author this text.

And my thanks to Shiella Patterson for her assistance in helping me complete this text on time and my cohort in law, Michael S. Schumann for his thoughts, ideas, and assistance.

THOMAS D. SCHNEID
Department of Loss Prevention and Safety
Eastern Kentucky University

CHAPTER 1

FUNCTIONS AND FRICTIONS

> We will tackle the giants and set them back because we have a better product and better know-how of the marketplace.
>
> —Thomas Casey

> One hand cannot applaud alone.
>
> —Arabian Proverb

The safety and resource control management function within today's industries is often a dichotomy of numerous identified and unidentified responsibilities and requirements assembled to achieve the ultimate purpose of safeguarding the entity's assets. Although the specific *job description* for the safety and resource control function may vary among and between organizations the goal of virtually all members of our profession is, first and foremost, to protect our human assets and, secondarily, to protect our tangible and intangible assets in a cost-effective manner. The goal is virtually universal in the profession and is a worthy objective for all organizations.

The problem that often arises in the safety and resource control function is when other goals and objectives of the organization directly or indirectly conflict with the goals of the safety and resource control function. Although virtually no company would advertise that they are abandoning a safety goal in pursue of another *priority* objective (e.g., increased production, increased staffing, and so on), this *re-*

prioritization happens all too frequently in today's business world. The safety and resource control function is not forgotten but simply gets "back-burnered" or reduced on the company's priority list of importance. In many cases, the necessary resources are reduced or eliminated, leaving the safety and resource control manager with all the responsibilities of protecting the assets of the company, with a reduced or eliminated capacity to manage the function effectively, if at all.

One of the more frequent modifications to the safety and resource control function that can also be a source of stress and potential conflict is the combination of separate responsibilities, which are often "lumped" by the organization under the umbrella of the safety and resource control position. Historically, the safety function involved primarily compliance with Occupational Safety and Health Administration (OSHA) regulations. In many organizations, the requirements to achieve and maintain compliance have required substantial resources and expertise. This one "hat" was often more than could be handled by individual managers; given the nature of the job function, however, upper management often neither saw nor appreciated the substantial nature of this necessary work. The safety function did not make any products, did not sell any products, was not a "revenue generator" for the company, and was often "foreign territory" for many managers who did not understand the value of the function. In fact, even the statistics often used by many companies to keep track of how well or poorly the company was doing in the safety area were conversely reversed from the usual methods and often diluted by other factors thus leading to misunderstanding and often mistrust of the data generated by the function. For example, production managers' performance are often evaluated on the number of widgets produced, as well as the quality of their work. The numbers are positive in nature. Whereas safety and resource control managers often evaluated on the number of injuries or illnesses that incurred, as well as the dollars spent on medical costs and other related costs, which is negative or reactive in nature. Thus, the safety function was often evaluated on the basis of the reactive results, with minimal consideration provided to the numerous influencing factors and situational factors outside of the function's control.

In addition, there was often a distinct lack of knowledge concerning the actual job activities of the safety function. Managers in the upper echelons of many organizations were aware that they needed to manage the function, yet they possessed a level of knowledge just sufficient to know to acquire the necessary expertise. Thus, when other requirements or functions moved up the list of priorities, such as environmental regulations, or other function responsibilities became vacant due to job movement or reprioritization, the safety and resource control function was often the "dumping ground" for these responsibilities. The terminology commonly heard around corporate America was to "do more with less," and the safety and resource control function was a viable area to dump related staff functions.

Another factor that often resulted in additional job responsibilities being housed within the safety and resource control function is the actual or perceived expertise of the individual. As one manager stated when reviewing new environmental regulations, "This stuff looks like the OSHA regulations. Give it to the safety department." Although the safety and resource control function often possesses correlating skills and abilities to manage related functions such as environmental, security, and human resources, the issue often becomes a matter of time. There are only 24 hours in a day and, with multiple responsibilities, the safety and resource manager usually must prioritize time to complete the requirements. In many cases, the manager becomes a "fire stopper" addressing the inevitable daily crisis permitting little or no time to effectively manage the overall safety and resource control function. Additionally, as is often the case, appropriate resources are not provided to correlate with the additional job responsibilities.

Of the major "frictions" often identified by safety and resource control practitioners in the field, the two primary areas involve lack of management commitment and lack of authority. The safety and resource control function is a support function, necessitating commitment from top-level management in terms of funding, personnel, and other resources as well as management support, in order to achieve success. This "management commitment" must be genuine and continuous and cannot be available only when it is "politically correct." Conversely, the lack of management commitment or management commitment that "ebbs and flows" with circumstances will not permit the safety and resource control function the ability to achieve a sustained and consistent program that will achieve the necessary results.

In addition, the lack of authority by the safety and resource control function is often linked to the lack of management commitment by upper management. With the safety and resource control function often being placed in a staff role within the management heirarchy, the safety and resource control function often "consults" with the management team to design, develop, and implement the necessary programs. All levels of the management team should be aware of the commitment by upper management (i.e., the boss) and work with the safety practitioner to achieve the goal. However, where management commitment is missing, the safety and resource control manager often becomes frustrated because of the actual or perceived lack of authority to motivate other members of the management team to achieve the safety and resource control-related goals and objectives.

One of the most recent "frictions" that has surfaced in may industries is the increase in job turnover. Although turnover was historically perceived as a human resource department problem, turnover has impacted production, quality as well as having a direct impact on the safety and resource control function. Turnover, especially today with downsizing, rightsizing, corporate takeovers, layoffs, and the rise of entrepreurism, can be especially high in some industries, which can result in increased accidents, increased workers' compensation costs, increased litigation costs, and other detrimental results that are usually beyond the control of the safety and resource control function.

The world of safety and resource control is undergoing significant changes virtually on a daily basis with the increased use of technology, the influence of the stock market on the way businesses are managed, and the driving influences on the safety area. Safety and resource control managers today possess numerous new ways of managing the function ranging from E-mail to videoconferencing. This new technology can greatly assist the safety function in achieving better efficiency and control of the numerous areas of responsibilities or can be a major "friction" for safety and resource control managers, depending on the ease of use, learning curve, and comfort level with this equipment. Safety and resource control managers usually possess minimal time to learn and re-learn new software and hardware thus the technology is often not used to its greatest potential.

The safety and resource control world, as with the business world in general, has also been greatly influenced by the major fluxuations and focus on our world's stock markets. The combination of instaneous information and worldwide markets, as well as the substantial increase in participation by "mom and pop" investors through retirement plans (e.g., 401 K plans, IRA, SEP), has transformed many companies from long-range planners to short-term planners. This fundamental shift in the way companies plan and react has significantly influenced the methods we use today to establish and maintain safety and resource control programs. Safety and resource control managers can no longer wait for the most opportune time to establish a program—the programs must be done now.

Safety and resource control has become a dollar-driven function, not unlike production, rather than a compliance driven function as in the past. The cost of workers' compensation, the cost of property insurance, liability insurance as well as the cost of litigation far exceeds the potential of monetary fines from the Occupational Safety and Health Administration (OSHA), the Environmental Protection Agency (EPA), and other governmental agencies. Companies today expect, if not demand, that the safety and resource control function not only prepare the organization for compliance with governmental regulations but also to minimize the potential risks and reduce the outflow of dollars in these areas.

The safety and resource control function today is significantly different than the past but still maintains the same basic objective of safeguarding the human assets of the organization. However, today's safety and resource control function is broader in scope, moves at a significantly faster pace, and possesses more components than we have ever experienced. Safety and resource control is everyone's responsibility, and the safety and resource control manager serves the leader of the function. Safety and resource control professionals must be prepared to accept this challenge and succeed on all fronts.

CHAPTER 2

LEARNING FROM THE PAST

> The reason so many people never get anywhere in life is because, when opportunity knocks, they are out in the backyard looking for four-leaf clovers.
>
> —Walter P. Chrysler

> He who learns but does not think is lost, he who thinks but does not learn is in danger.
>
> —Confucius

In terms of the history, safety and resource control-related laws and regulations are relatively new; many are less than 30 years old. However, during this relatively short period, a virtual explosion of information, philosophies, and programs has evolved to address identified hazards or risk of loss situations. C. Everett Marcum, Frank Byrd, Dan Petersen, and many others have blazed a a pathway through the regulatory forest, providing the basis from which we can build the future of safety and resource control management.

In the broadest of terms, what can we learn from the past triumphs and failure? Consider what we have learned:

- We can manage this function.
- Safety and resource control programs save lives and save money.

- Success in safety and resource control requires management commitment and the correlating resources.
- A myriad of regulations govern the safety and resource function (e.g., OSHA, EPA). Compliance does not automatically equal success.
- Accidents can be prevented.
- If our programs are not in written form, we don't have a program.
- There is no such thing as "bad luck"—only bad management.
- Success requires expertise, diligence, and consistency.
- The function requires good communications skills and training skills.
- The function requires careful and thoughtful planning.
- The function can be successfully managed.
- Success can be achieved in effectively and efficiently managing the safety and resource control function.

Above all, the commitment of management to the safety and resource control efforts is essential for success. Without the support, personnel, and funding, safety and resource control program will be nothing but window dressing and will cause more harm than no program at all. Under most circumstances, when management fully understands the cost factors involved in work-related accidents, the safety and resource control manager will begin to achieve the necessary "buy-in" to achieve success. Safety and resource control managers should be prepared to show the results, in both monetary and humanitarian terms, that can be acquired through a comprehensive and systematic management approach to safety, health, loss prevention, and other resource control functions. To ensure that management fully understands the concepts involved in a proactive program, all levels of the management team should understand how accidents happen and how accidents can be prevented. Using the basic domino theory, safety and resource control mangers can easily explain the causal factors that lead up to a accident and the negative impact after an accident. In addition, safety and resource control managers can explain that, through the use of a proactive safety, health, and loss prevention program, the factors that could lead to an accident can be identified and corrected before risk factors mount and ultimately lead to an accident.

Focusing on the methodology used in the management of safety and resource control, several management theories and approaches, including, but not limited to, Management Control System Management,[1] Management by Objectives,[2] Group Dynamic and Human Approach Management,[3] and Total Safety Management,[4] have successfully been used in different organizations (see Chapter 8). The particular management theory selected for use within any given organization must meet the needs and management style of the organization. There is no one right or wrong management theory for any given organization as long as the management system selected provides a consistent, systematic approach that proactively addresses the underlying reasons and risk factors that may ultimately lead to an accident.

Many organizations have found that the Management By Objectives (MBO) theory is a simplistic but effective systematic and practical methodology for the management of their safety, health and resource control function. This style provides a stair-step long-term approach to achieving the ultimate safety goals or objectives. Using the MBO theory, each element within a safety or resource control program can be assigned an achievable objective or goal. When all objectives from each elements within a specific program are achieved, the overall objective of the program will be achieved concurrently. When all individual safety and resource control program objectives are achieved, the larger overall objective of safety and resource control effort will be achieved. In simple terms, MBO provides a series of building block objectives on which other objectives are based, achievement of the smaller objectives will ultimately lead to the achievement of the larger objectives or goals.

In developing specific safety, health, loss prevention, or resource control goals or objectives for an organization, all levels of management and employees should be provided an opportunity to interject their ideas and opinions in the development of the safety, health, loss prevention or resource control goals or objectives. There are two basic schools of thought in this area: *zero accident/loss goal theory* versus the *progressive accident/loss goal theory*. Under the zero accident/loss goal theory, the ultimate goal is zero accidents or zero losses. To attain less than this goal is to permit employees to incur injuries and illnesses on the job or losses to be incurred. Using the ultimate goal of zero, the entire organizational team possesses a common goal that is the pinnacle of the safety, health, loss prevention, or resource control summit. The downside of the zero accident/loss goal theory is the possibility that organizational team members may view the zero accident/loss goal as unrealistic and unattainable and thus lose interest and momentum in striving to achieve their safety and resource control goals. Under the progressive goal theory, organizations will continuously phase in reducing safety, health, loss prevention, or resource control goals over a period of time, in an effort to achieve the ultimate goal of zero accidents (e.g., 1991, 25% reduction from 1990 accident total; 1992, 50% reduction from 1990 accident total, ultimately reaching the zero accident goal over a number of months or years). The downside of the progressive accident/loss goal theory is the fact that the organization will be accepting a certain number of accidents or losses, and thus injuries and illnesses, while the organization strives to achieve the ultimate goal.

Although the involvement of every employee of the company is important in any safety and resource control program, the key management position level is the first-line supervisor or team leader. This management level is normally the communications link between upper management and the employees and often serves as the personnel and human resource managers "eyes and ears" in the production areas. In most organizations, the first-line supervisor or team leader is the person who will interact on a daily basis with the employees within his or her department or area, direct the activities of the employees in the department or area, proscribe or perform disciplinary functions, and/or perform the training function and other related activities. This management level embodies the commitment of the organization to the safety, health, and loss prevention and relays it to the employees. If the first-line supervisors or team leaders have been properly educated in, and adopt, the goals and objectives of the safety, health, loss prevention, or resource control program, and effectively communicate these goals and objectives to their employees, the employees will normally embrace the safety and loss prevention effort or, at the very least, adhere to the safety, health, loss prevention, or resource control policies and procedures.

First-line supervisors or team leaders should be educated, trained, and motivated to make safety and resource control part and parcel of their everyday activities. First-line supervisors and team leaders must be provided the tools from which they can effectively manage the safety, health, loss prevention, and resource control function just as they manage production, quality, and other job requirements. Commitment and motivation on the part of upper-level management, in combination with the necessary education and training (i.e., the tools), for supervisors or team leaders to manage safety and resource control effectively, as well as holding the supervisor or team leader accountable for the safety and resource control performance or achievement of the goals or objectives, normally prove successful in acquiring the appropriate buy-in.

One of the first questions normally asked by first-line supervisors and team leaders is, "Where am I going to find the time to manage the safety and resource control function when I don't have enough hours in the day to complete all my job responsibilities now?" The proactive approach to safety and resource control provides the first-line supervisor or team leader with the skills to manage the safety and resource control function within their department or area effectively, instead of reacting to problems. The management skills taught for the effective management of the safety and resource control function are the same basic management skills necessary to manage the production function, quality function, and other related functions effectively. Supervisors and team leaders normally find that when they have mastered the basic management skills, the safety and resource control function can be managed effectively in the same or similar manner as the other production functions and, in fact, the supervisor or team leader will acquire more time within the workday when he or she manages, rather than putting out fires.

The management principles used by management team members in the daily supervision of production, quality control, or any other operation are the same when managing safety in the workplace. In production, the supervisor plans, organizes, directs, and controls his/her operation to produce a product, whereas in safety and loss prevention, the supervisor plans, organizes, directs, and controls[5] the safety and resource control function in the workplace. Basic management skills used in production and quality are transferable to the safety and resource control function.

Another area that normally requires substantial effort in managing the safety and resource control function is achieving an open communication system with employees. All employees want to be able to work safely, to avoid injury while at work, and to control losses in their job. The goal of management is the same. The confrontation normally occurs in the methods used to achieve this uniform goal. Communication with employees, permitting employees to voice their opinions and ideas, and acquiring employee involvement in the safety and resource control effort are essential to the proper management of the safety and resource control program.

One of the cornerstones of most safety and resource control programs is the presence of written programs, policies, and procedures through which the organizational team members, individually or collectively, can acquire the necessary guidance regarding acceptable and unacceptable behaviors, expectations as to safety and resource control performance, and other basic workplace requirements. Safety and resource control policies and procedures should be clearly stated, with any ambiguities or room for interpretation removed. Written safety and resource control programs provide the essential requirements of the specific safety and resource control program, which is vital in providing continuous direction. There is no perfect safety and resource control objective or goal mechanism that works for all organizations. Given the substantial differences in location, worksite, workforce, and philosophy, safety and resource control managers should select the mechanism or method that works best for their individualized situation. The key factors in safety and resource control program development under this management theory are that (1) the organizational team possesses a consensus safety and resource control goal; (2) the objectives in attaining the goal are clearly defined and measured; (3) the organizational team is provided input as to their achievement of the safety and resource control objectives and goals; and (4) the organizational team is held accountable for the achievement of the safety and resource control goals.

In developing a written safety and resource control program, there is no substitute for knowledge of the OSHA standards, EPA regulations, or other applicable government regulations. Under the law, every organization covered under these regulations is bound to know the law. As stated by many courts throughout history, ignorance of the law is no defense.

A basic and general guideline to assist the safety and resource control managers in developing a safety and resource control program for a particular standard is set forth below:

1. Read the OSHA standard or regulation carefully, and note all requirements.
2. Remember that most OSHA compliance programs or governmental compliance programs must be in writing.
3. Develop a plan of action. Acquire management commitment and funding for the program.
4. Purchase all necessary equipment. Acquire all necessary certifications.
5. Remember to post any required notices.
6. Inform employees of the program. Acquire employee input in the developmental stages of the program. Inform labor organization, if applicable.
7. At this point, you may want to contact OSHA or another government agency (or your network contacts) for samples of acceptable programs (they sometimes have available a recommended format). You may also want to have them review your finished draft and provide comments.
8. Conduct all necessary training and education. Remember to document all training.
9. Conduct all required testing. Remember to document all testing procedures, equipment, calibrations, and so forth.
10. Implement the program.
11. Ensure that all procedures are followed. Disciplinary action taken for noncompliance must be documented.
12. Audit the program on a regular periodic basis, or as required under the standard.

Simply complying with the OSHA standards or other government regulation does not guarantee successful safety and resource control program. Most OSHA standards or other government regulations are the bare bones, or minimum requirement, that the government requires all employers to meet. A safety and resource control program must comply with these standards but should go far beyond the minimum standard. A good program should incorporate ideas and programs developed by the employees and management team to strengthen and expand the safety and resource control efforts. Many of the best ideas in the safety and resource control area have been originated by employees. Safety and resource control managers should bear in mind that employees normally work in one area and perform one job. The employees are the expert on that particular job; their ideas and input can normally provide great insight into developing safety and resource control programs and policies that have direct effect on that particular job or area.

The basic concept in managing safety and resource control in the workplace is to get all employees to be conscious of their own safety as well as the safety of others. Safety and resource control can be instilled into employees through a long-term training and education program and by constant, consistent, and proper management of the safety and resource control function. Safety and resource control should be made an essential part of each employee's daily work habits. Employee involvement in the structure, decision making, and operation of the proactive safety and resource control program has often proved successful in achieving employee "but-in," and thus the commitment. Safety and resource control is not the sole domain of the safety director, personnel manager, or even the first-line supervisor. The team approach permits the supervisor or team leader to train organizational team members to take an active role in the specific safety and resource control functions. Many organizations have found that safety and resource control activities required for the achievement of specific objectives, such as department safety inspections, personal protective equipment inspections, and other duties, can be delegated from the first-line supervisory level to the team members. In fact, the more involved the organizational team members are in the safety and resource control program, the more they feel responsible for the safety and resource control program. However, too much delegation of essential duties can defeat a good program.

Another key area that is often overlooked in the management of a safety and resource control program is the accountability factor. All levels of the management team must be held accountable for their divisions, departments, or area. The individual management team member should be involved in the development of the objectives and goals, as well as the necessary tools, to enable the management team member to manage the safety and resource control function effectively. Pertinent and timely feedback is critical.

The use of positive reinforcement has been found to be the most effective method in motivating supervisors or team leaders to achieve the specified objectives and goals. However, negative reinforcement or disciplinary action should be in place as a backup if positive reinforcement is not successful. Safety and resource control managers should ensure that a fair and consistent policy with regard to disciplinary action in the area of safety and resource control is established and maintained. Organizations that have embraced the proactive approach to managing the safety and resource control function have found that the benefits achieved over time far outweigh the initial costs involved and, once in place, an effectively managed safety and resource control program will pay dividends for years to come, as well as minimize potential risks and potential legal liabilities.

The ultimate goal for every safety and resource control manager is to safeguard the organizations assets from harm in the workplace. A secondary goal that is vitally important is the achievement and maintenance of compliance with the government standards and requirements. In order to reach these important goals, a comprehensive management approach should be developed to manage the safety and resource control and an extensive, all-inclusive strategy through that to direct and control the completion of the required tasks in order to achieve compliance with the government standards and regulations.

The principles used by managers in their daily supervision of production, quality control, or any operation are the same that should be used when managing the safety, health, and loss prevention function in the workplace. In production, managers use the basic management principles of planning, organizing, directing, and controlling[6] the operation to produce a product. Safety and resource control managers can use and teach the same basic management principles to plan, organize, direct, and control the safety and loss prevention function in the workplace. With the increasing costs of work-related injuries and illnesses, the increasing compliance requirements and liability, increasing risks and losses in the workplace, as well as other increasing costs in the area of safety and resource control, a proactive management approach has proved the most effective way of ensuring a safe and healthful environment is created and maintained in the workplace.

ENDNOTES

1. Bird and Germain, *Practical Loss Control Leadership*, ILCI Press, p. 41 (1990).
2. C.E. Marcum, *Modern Safety Management*, Kingsport Press, 1978, p. 29.
3. D. Petersen, *Safety Management: A Human Approach*, Aloray, Inc., pp. 9–23.
4. Total Safety Management is a derivative of the Total Quality Management approach.
5. *Managing Employee Safety and Health Manual*, Tel-A-Train, Inc. (1991).
6. Ibid.

CHAPTER 3

DRIVING FORCES

> Change is no modern invention. It is as old as time and as unlikely to disappear. It has always to be counted on as of the essence of human experience.
>
> —James Rowland Angell

> Change is inevitable, except from a vending machine.
>
> —Bumper Sticker

The safety and resource control function, and its various components, has and is undergoing significant changes over the past 30 years. In essence, the modern age of safety started with the inception of the Occupational Safety and Health (OSH) Act of 1970; companies acquired professionals with expertise primarily to manage the compliance function. The driving force to integrate safety and resource control into the management function for many companies during the 1970s was simply to achieve and maintain compliance with Occupational Safety and Health Administration (OSHA) standards. In general, companies wanted to avoid the monetary penalties that OSHA could issue for noncompliance and to create a safer work environment for their employees.

However, over the years, many companies and organizations found that there was a side benefit to compliance in the fact that fewer injuries and illnesses equated to fewer workers' compensation dollars being spent that directly or indirectly positively affected the bottom line. In addition, during the 1970s and 1980s, many state legislatures increased the benefit levels for workers' compensation, which steadily drove workers' compensation to become one of the driving forces for safety and resource control management.

Throughout this period, another factor emerged as a primary reason or driving force for safety and resource control management, namely litigation. Our American society has become increasingly litigious. Companies are being sued virtually on a daily basis for their work practices and activities, discrimination in the workplace, product liability, and numerous other causes of action. Win, lose, or draw, companies are spending billions of dollars defending their products and actions. American industry has emerged as a major target for litigation because of their deep pockets and often willingness to settle issues to avoid the potential of astronomical jury awards and bad publicity. The primary method of avoiding litigation is to adopt a proactive program to identify areas of potential litigation (e.g., discrimination in the workplace) and develop policies, procedures, and programs to eliminate and/or minimize the potential risk.

The safety and resource control function has also become a prime player in the area of labor organization avoidance. Historically, employees have joined or formed unions in order to engage in collective bargaining with their employer from a position of power regarding issues of wages, hours, and conditions of employment. A key issue in many union-organizing campaigns or a primary concern of the employees is working in a safe and healthful environment. Employers have learned that employees will not tolerate unsafe and unhealthful work environments today as their grandfathers and great-grandfathers did in the past. Where these conditions exist, employees have many avenues of redress, including forming and joining a union. Employers have learned that where employees are provided a good wage and reasonable benefits, work appropriate hours, and have a safe and healthful workplace, employees generally have no need to join or form a union. Conversely, where working conditions are unsafe or unhealthy and the employer does not address these important issues, employees will not tolerate these conditions, and the potential of the employees organizing is high. The safety and resource control function creates and maintains a safe and healthful work environment as part and parcel of eliminating or minimizing employees need for collective bargaining.

Today's technology has also created a new driving force for many companies. What comes to your mind with the words "Bhopal, India" or "Exxon Valdez?" Many companies today are more image conscious and definitely more shareholder conscious. Companies often spend millions of dollars on Madison Avenue to establish their name and products. Companies also closely monitor the views of the owners of the companies, that is, the shareholders. Given the instantaneous television coverage and information exchange via the Internet and other sources, a "bad accident" at a company can result in coverage on the news networks within an hour with a downstream affect not only on the company's image but also on its value, namely the stock price. There is no "upside" to a major catastrophe in a company. In many scenarios, the company's image can be tarnished, causing the public to avoid purchasing their products. Shareholders would sell their stock, reducing the value of the company, and detrimentally affecting their financial position. OSHA and "everyone who is anyone" would descend upon the company, causing long-term disruption as well as possible monetary penalties. In short, corporate executives have seen these catastrophes happen and they want to avoid the catastrophe at their company. The method of avoiding the potential of a catastrophe is through a proactive and efficient safety and resource control management effort.

As we move into the new millennium, OSHA compliance, reduction of workers' compensation, avoidance of litigation, and catastrophe avoidance continue to be driving forces in safety and resource control profession; however, new areas are quickly emerging or reemerging. OSHA compliance is now delving into more difficult areas such as workplace violence and tuberculosis, creating a renewed emphasis on compliance. Legislatures are modifying state workers' compensation acts to address issues such as cumulative trauma disorders and workplace stress. Litigation is expanding to encompass entire industries such as the tobacco and firearms litigations. Shareholders are now actively involved in their investments in companies and are trading online. Our workforce is more sophisticated, more mobile, aging and fewer in numbers. Acquisition and retention of the workforce is emerging as a driving force. Our workforce is using more brain power than muscle power. Our workforce expects not only a safe workplace for themselves, but also safe day-care for their children and safe elder care for their parents. Our workplace is changing and our driving forces will change, ... but our main objective will remain the same.

CHAPTER 4

IN SEARCH OF THE SAFETY GRAIL

> I have wandered all my life, and I have also traveled; the difference between the two being this, that we wander for distraction, but we travel for fulfillment.
>
> —Hilaire Belloc

> Lots of folks confuse bad management with destiny.
>
> —Kin Hubbard

Safety and resource control managers wear many hats and often get bogged down in the details of everyday activity. Safety and resource control managers often look at their "to do" lists at the end of the day and find that very few of the important items scheduled for attention were actually initiated or completed because of the numerous interruptions, unexpected "fires" that needed to be extinguished, and other unplanned activities. Over time, this constant barrage of activities and playing catchup on scheduled activities can engulf the safety and resource control manager. In essence, the safety and resource control manager is struggling to survive in the forest and often can't see the trees.

On a periodic basis, it is important for safety and resource control managers to simply step back and assess and analyze where they are in their careers and development, as well as what is going on with their program. In our search for the ultimate safety grail, the search and growth achieved as the result of the struggles and efforts is often more important than the actual achievement of the ultimate goal.

The safety and resource control function is significantly different from other managerial functions within most operations. The ultimate grail for production is making the perfect widget, for quality control it is making the consistently perfect widget, and for engineering the grail is designing of the operation to make the most cost-efficient widget. In the safety and resource control function, the grail consists of preventing or protecting assets, whether human or otherwise.

The safety and resource control function usually establishes long-range goals or objectives to guide our search for the grail. To establish these long-range goals and objectives, safety and resource control professionals should dream or visualize of the anticipated pathway to achieve the grail. A virtual road map should be formulated, identifying the best routes, anticipated battles, areas of alliance, and other factors and issues that may be encountered in the quest for the grail. Safety and resource control professionals can call upon their past experiences and successes and failures, as well as the ideas and programs of others, but they should also "think outside of the box" to create new and revolutionary ideas.

Many safety and resource control manager strives daily to achieve the grail while addressing the day-to-day scheduled and unscheduled activities and faced with the uncertainty that there are many factors beyond his/her control that have a detrimental impact on the achievement of the goals or objectives. Safety and resource control managers should be aware that their daily battles can become cumulative in nature, with the manager becoming consumed with the battle and losing sight of the grail.

To ensure that safety and resource control crusaders remain true to the search, they should allot time on a periodic basis to pull the map out and make sure they are on the right road. Periodic assessment of your safety and resource control goals and objectives, your time line for achievement of the objectives, your anticipated obstacles, and your "battle plan" is essential.

Keeping the grail within long-range focus often requires scheduling a time in which the safety and resource control manager can emerge from the forest and see the pinnacle of the grail. When the grail is again seen and back in focus, it is often important for the safety and resource control manager to adjust and reassess the short-term goals and objectives, the time line and other components of the "battle plan" to maintain the truest path to the achievement of the grail. This assessment and adjustment must be a periodic planned activity or the safety and resource control manager risks wandering through the forest in directions unknown expending energy, resources, and time in a fruitless search.

To assist safety and resource control managers in their search, the following guide will assist in the development of your map:

My grail is: ____________________

My path to my achievement of my grail includes the following goals or objectives: ____________________

____

My path will be strewn with the following obstacles: ____________________

____

I will need the following "tools" in my quest: ____________________

____

I will need the following team members to assist me on my quest: ____________________

____

My "map" to guide my quest includes: ____________________

I will achieve my first goal or objective by: ____________________

I will achieve my second goal or objective by: ____________________

I will seize my grail no later than: ____________________

Managing Your Future Career: Dream Weavers
By Mark D. Hansen, CSP, CPE, PE

The most successful business people I have met in my career are dream weavers. They supplant a "bone-deep belief" that, there are people who do agree and will pursue them with their dream and vision. This energizes those who work with and for dream weavers. Martin Luther King did not galvanize a whole race by saying, "I have an objective." He did it by saying, "I have a dream." Everyone bought into that dream and the rest is history. Dream weavers cultivate that same kind of emotion in people. Further, it means that their success does not necessarily mean failure for others, just as their success does not preclude my own.

All too often, we gauge our success by someone else failing. This ultimately sets up a win–lose or a lose–lose situation where we end up with Phyrric victories. That is, we win but the cost of the battle was too devastating to both parties.

I have observed that dream weavers often make the difference between excellence and mediocrity, particularly because they virtually eliminate small thinking and adversarial relations. What characteristics differentiate dream weavers from others? Consider the following.

They return often to the right source. Dream weavers drink deeply from sources of internal security—sources that renew and recreate them. Those internal sources nurture and nourish their soul, enabling them to grow, and receive comfort, insight, inspiration, protection, direction and peace of mind. They seek solitude and enjoy quiet time. Dream weavers practice reflection and contemplation. They find opportunities to be alone and to think deeply, to enjoy silence and solitude. They make time to reflect, write, listen, plan, prepare, visualize, ponder, and relax.

They sharpen their saw regularly. Dream weavers cultivate the habit of "sharpening the saw" every day by exercising mind and body. It means that they are strong enough, intelligent enough and confident enough to take on opposing views and discuss them openly.

Source: From Class Edition, ASSE, Winter 1999, 1.

They serve others. By returning often to nurturing sources of internal security, dream weavers restore their willingness to serve others effectively. In safety, we are to serve our employers and assist eliminating injuries and illnesses. There is a measure of humility with this type of job. If we are not here to serve, it becomes readily evident. If we lack humility others can read through us quickly. Dream Weavers take particular delight in anonymous service, feeling that service is the rent we pay to live in this world. If our intent is to serve others without self concern, we are rewarded with increased internal security.

They lead balanced lives. They keep up with current affairs and events. They read a variety of books. They are active socially and intellectually. Their sense of their own worth is manifested by their courage and integrity and by the absence of a need to brag, to drop names, to borrow strength from possessions or credentials or titles of past achievements. They maintain a long-term intimate relationship with another person.

They believe in other people. Dream weavers don't overreact to negative behaviors, criticism or human weaknesses. Rather they look inward to correct problems. They are not quick to blame but rather they are accountable for their errors. They are mentors to others without any requirement for a return on their investment, because it usually pays big dividends just to see the growth in those being mentored. They see the oak tree in the acorn and understand the natural process by which they can help this acorn become a great oak.

They radiate positive energy. Dream weavers are cheerful, pleasant, and happy. Their attitude is optimistic, positive, upbeat. Their spirit is enthusiastic, hopeful and believing. This positive energy surrounds them like an aura, attracts and charges people to want to be around them. Everyone wants to be part of a success story and conversely, no one wants to be part of a failure. Dream weavers savor life and its adventures. Their security lies in their initiative, resourcefulness, willpower, courage, stamina, and native intelligence rather than in the artificial safety of their home camps and comfort zones. They are pushing the envelope to make the most of every event.

They are synergistic. They are change catalysts. They champion the improvement of almost any situation they get into. They work as smart as they work hard. And they are amazingly productive.

Are you a dream weaver or a dream weevil? Success is what you make of the situation you are in. 99% of our success is based on our attitude in how we react to what happens to us. We can't control what happens to us, but we can control how we react to it. Take the time to stop and smell the roses, think, reflect, and ponder.

CHAPTER 5

ONE SIZE DOES NOT FIT ALL

> The block of granite which was an obstacle in the path of the weak, becomes a steppingstone in the path of the strong.
>
> —Thomas Carlyle

> Nothing in life is to be feared.It is only to be understood.
>
> —Marie Curie

A frequently encountered mistake made by small employers and new safety and resource control professionals is the use of standardized programs or "borrowed" programs from the predecessor or other professionals. Although there is no prohibition against the use of such programs, they may be correct in structure and content yet fail to fit the individual company or operation. In addition, many safety and resource control professionals have found that programs that work in one facility or operation may not work in another facility or operation. Virtually all programs, whether compliance-oriented or otherwise, must be customized to "fit" the employees and operations of the given area.

In the compliance area, the initial foundational elements are usually provided within the Occupational Safety and Health Administration (OSHA) standard or other government regulation. However, most government regulations are not written in a format that informs

BAD

how to achieve compliance but in a format that instructs the minimum level of compliance required to achieve compliance. The OSHA standards and other regulations inform the safety and resource control professional, in essence, that you are at level A and must achieve level Z. Most standards and regulations do not provide guidance as to how to achieve Z only that Z must be achieved in order to be in compliance. How the safety and resource control professional achieves Z is usually a judgment call on the part of individual professionals based on their expertise, management system, experience, and many other factors.

The foundational basis necessary to manage a proactive and effective safety and resource control program properly begins with the presence of customized written safety and resource control policies and procedures through which the organizational team members, individually or collectively, can acquire the necessary guidance regarding acceptable and unacceptable behaviors, expectations as to safety and resource control performance, and other basic workplace requirements. Safety and resource control policies and procedures should be clearly stated while removing any ambiguities or room for interpretation. In addition, as discussed in detail later in this chapter, detailed written safety and resource control programs providing the essential requirements of the specific safety and resource control program (based on the Z required by the standard or regulation) is essential in providing continuous direction and guidance to the organizational team. Given the substantial differences in location, worksite, workforce, and philosophy, safety and resource control professionals should carefully select the mechanism or method that works best for their individualized situation. There are several key factors in safety and resource control program development: (1) the organizational team should possess a consensus safety and resource control goal or grail; (2) the objectives in attaining the goal are clearly defined and measured; (3) the organizational team is provided input as to their achievement of the safety and resource control grail; and (4) the organizational team is held accountable for the achievement of the safety and resource control objectives.

Another area that normally requires substantial effort in managing the safety and resource control function is achieving an open communication system with employees. All employees want to be able to work safely and not be injured while at work. The grail of the management team, in whole or in part, is often the same. Confrontation can occur in the methods used to achieve this jointly held objective. Communication with employees, permitting employees to voice their opinions and ideas, and acquiring employee involvement in the safety and resource control effort are essential to the proper management of the safety and resource control program.

Remember that simply complying with the OSHA standards or most government regulations does not guarantee a successful safety and resource control program. The OSHA standards and other government regulations are often the bare bones minimum requirement set forth by the government for all employers in virtually all industries to meet. A safety and resource control program must comply with these standards but should go far beyond these minimum standards. A good program should incorporate ideas and programs developed by the employees and management team to strengthen and expand the safety and resource control efforts. Many of the best ideas in the safety and resource control area have originated with employees. Safety and resource control professionals should bear in mind that employees normally work in one area and perform one job. The employees quickly become the resident expert with regard to that particular job; their ideas and input can normally provide great insight into developing safety and resource control programs and policies that directly affect that particular job or area.

A basic and general guideline to assist the safety and resource control professionals develop a safety and resource control program for a particular standard or regulation is set forth below:

1. Read the OSHA standard or agency regulation carefully and note all requirements.
2. Remember most OSHA compliance programs or agency regulatory programs must be in writing.
3. Develop plan of action. Acquire management commitment and funding for the program.
4. Purchase all necessary equipment. Acquire all necessary certifications.
5. Remember to post any required notices.
6. Inform employees of the program. Acquire employee input in

the developmental stages of the program. Inform labor organization, if applicable.

7. At this point, you may want to contact OSHA or the agency for samples of acceptable programs (they sometimes have available a recommended format). You may also want to have them review your finished draft and provide comments.
8. Conduct all necessary training and education. Remember to document all training.
9. Conduct all required testing. Remember to document all testing procedures, equipment, and calibrations.
10. Implement the program.
11. Remember to insure that all procedures are followed. Disciplinary action taken for noncompliance must be documented.
12. Audit the program on a regular periodic basis or as required under the standard.

Again, it is important to note that most OSHA standards and other governmental regulations do not provide any guidance as to how compliance is to be achieved. Most, if not all, standards and regulations require only that the employer achieve and maintain compliance. In essence, the OSHA standard or other regulations tell the safety and resource control professional what has to be done but not how the compliance program is to be designed, managed, or evaluated. For example, an employer must have in place a facility evacuation plan. How the employer structures the plan and the specific details of the plan is left to the safety and resource control professional. It is the responsibility of the safety and resource control professional to determine what OSHA standards are applicable to the individual facility or workday to ensure that the facility and workday are in compliance with the applicable standards. An error of omission in not possessing a part or all of the required safety or other compliance program necessary to achieve compliance with a specific standard is a violation, just as an inadequate or mismanaged program is a violation. Errors of omission and commission are violations of the OSHA Act and most government regulatory schemes.

The basic concept in managing safety and resource control functions in the workplace is to be able to get all employees to be conscious of their own safety as well as the safety of others. Safety and resource control can be instilled into employees through a long-term training and education program and constant, consistent, and proper management of the safety and resource control function. Safety and resource control must be made an essential part of each employee's daily work habits. Employee involvement in the structure, decision making, and operation of the practices of the safety and resource control program has often been found to be successful in achieving employee "buy-in" and thus the commitment. Safety and resource control is not the sole domain of the safety director, personnel manager, or even the first-line supervisor. Using a team approach, the supervisor or team leader can train organizational team members to take an active role in the specific safety and resource control functions. Many organizations have found that safety and resource control activities required for the achievement of specific objectives, such as department safety inspections, personal protective equipment inspections, and other duties, can be delegated from the first-line supervisory level to the team members. In fact, the more involved the organizational team members can be in the safety and resource control program, the more organizational team members feel responsible for, and take possession of, the safety resource control program. However, too much delegation of essential duties can defeat a good program. Another key area that is often overlooked in the management of a safety and resource control program is the accountability factor. All levels of the management team must be held accountable for their divisions, departments, or area. The individual management team member should be involved in the development of the objectives and goals, as well as the necessary "tools" to enable the management team member to manage the safety and resource control function effectively. Pertinent and timely feedback is critical.

The use of positive reinforcement has been found to be the most effective method in motivating supervisors or team leaders to achieve the specified objectives and goals which ultimately lead to the achievement of the grail. However, negative reinforcement or disciplinary action should be in place as a backup measure, in the event that positive reinforcement is not successful. Organizations that have embraced the practice approach to managing safety and resource control have found that the benefits achieved over time far outweigh the initial costs involved and, once in place, an effectively managed safety and resource control program will pay dividends for years to come as well as minimize potential risks and potential legal liabilities.

As discussed in Chapter 4, the ultimate grail, or portion thereof, for most safety and resource control professionals is to safeguard employees from harm in the workplace. A secondary goal, although vitally important, is the achievement and maintenance of compliance with the OSHA and related government standards and requirements. In order to reach these important goals, a comprehensive management approach needs to be developed to manage the safety and resource control function and an extensive all-inclusive strategy through that to direct and control the completion of the required tasks in order to achieve compliance with the OSHA and other government standards and regulations.

A management philosophy should be incorporated to serve as the foundation and to provide the necessary style through which to manage the safety and resource control function. The selection of that management philosophy and style is an individual decision based on the background and personality of the safety and resource control manager, type of industry, employee population, and numerous other factors. The key to achieving compliance with the OSHA or other government standards and regulations is to proactively manage the safety and resource control function, rather than permitting the safety or related issues or problems to dictate to the organization.

The management principles used by most management team members in the daily supervision of production, quality control, or any other operation are the same when managing safety in the workplace. In production, the supervisor would plan, organize, direct, and control the operation to produce a product while in safety and resource control, you would plan, organize, direct, and control[1] the safety, health, security, and related functions in the workplace.

The foundational issues that should be addressed include the following:

1. Is the employer covered under the OSHA Act or related regulatory agency?
2. Has the facility or workday been evaluated to ascertain which specific OSHA standards or other government requirements are applicable?

3. Has the management group been educated as to the requirements of the OSHA Act and standards and acquired the necessary support and funding for the programs?
4. Is there a copy of the OSHA standards (29 C.F.R. 1910 et seq. and other standards) or other government regulations at the work site?
5. Is the Federal Register or other appropriate sources for new standards or emergency standards being reviewed to see whether they are applicable to the workday?
6. For new programs or new standards promulgated by OSHA or other governmental agencies, is the standard or regulation applicable to the workplace, situation, or industry?
7. Is there an OSHA guideline or other guidance information or research for particular situations or hazards in the workplace?
8. If there is no applicable OSHA standard, will the situation qualify under the General Duty Clause as an unsafe or unhealthful situation or hazard? Other catchall clauses in other regulations?
9. If there is no applicable OSHA standard and the situation is deemed to qualify under the general duty clause, have National Institute of Occupational Safety and Health (NIOSH) publications, American National Standards Institute (ANSI) standards, or other applicable journals and texts been reviewed to acquire guidance?
10. If there is no applicable OSHA standard or there is an OSHA standard that conflicts with other governmental agency regulations, there are other options: (1) contact the regional OSHA office and request consultation assistance or; (2) employ an outside consultant possessing the specific expertise to assist; or (3) pursue a variance action?
11. If there is an applicable OSHA standard or other government agency regulation has each and every word of the standard or regulation been read, so as to completely understand the requirement of the standard or regulation?
12. Is there a written program to ensure compliance with all requirements of the OSHA standard or government regulation? Remember, if the program is not in writing, there is no evi-

dence to prove the existence of the program during either agency inspection or litigation. Is the original document for this written program kept in a secure location?

13. Is the program written in a defensive manner? Can the written program be scrutinized by OSHA or a court of law without identifying flaws in the program? Is the program written in a neutral and nondiscriminatory language?
14. Is the documentation of every purchase, equipment modification included in the program in order to ensure compliance with the OSHA standard or governmental agency regulation? Is this documentation in the original copy of the written program or in a secure location?
15. Does the written program possess the purpose of this program? Is there a copy of the applicable OSHA standard or governmental agency regulation for easy reference? Have the responsibilities been delineated? Are they specific for each level of the management team and each position within the levels of management?
16. Has the OSHA standards and other applicable information been closely scrutinized and evaluated? Has each and every requirement in the standard been achieved or exceeded? Have any steps or elements required in the OSHA standard or government agency regulation been omitted? If the standard is vague, make sure the program is clear, concise, and to the point.
17. Has all training been documented as required under the OSHA standard or agency regulation? (Remember, the OSHA standard and most government agency regulations are only a minimum requirement. Every program can be better than, but it must not be less than, the OSHA standard requirements or government agency requirements.)
18. Is there detailed documentation regarding each and every phase of training? Is this documentation in the written program? Is documentation provided in the written program to prove the use of audiovisual aids, the instructor's qualifications, and other pertinent information?
19. Is there a schedule of the classroom and hands-on training sessions in the written program?

20. Have all employees who have completed the required training signed a document showing the exact training completed (in detail), the instructor's name, and the date of the training? Have auxiliary aids and other accommodations for individuals with disabilities been provided for in the training programs? (For employees who cannot read/write, a thumb print can be used. Videotaped documentation of the training is also an acceptable method of documentation. Remember to maintain the individual tapes on file, as with other documents, for future use as evidence of this training.)
21. Is the training offered in the languages used by the employees? Are the documents and the written program interpreted into the languages spoken by the employees and are the interpreted programs provided for use by these employees?
22. Is there a posting requirement under the applicable OSHA standard or agency regulation (e.g., Wage and Hour)? Is the necessary poster from the governmental agency in the facility and has it been posted in an appropriated location by the required date? Have posters in the language of the employees been acquired and are they posted in an appropriated location by the required date?
23. Are there any other requirements? Labeling of containers? Material Safety Data Sheets (MSDS)? Does the OSHA standard or agency regulation require support or information from an outside vendor or agency? Have all requests to outside vendors or agencies requesting the necessary information (i.e., MSDS sheets) been documented and placed in the written program?
24. Is there a disciplinary procedure in the written program instructing employees of the potential disciplinary action for failure to follow the written program?
25. Has the written program been reviewed before publication? Does it meet and/or exceed each and every step required under the applicable OSHA standard or agency regulation?
26. Has the written program been reviewed by legal counsel or the upper management group prior to publication? Upon completion of the acquisition of necessary approvals of the written program, have copies of the program been made for distribu-

tion to strategic locations in the facility? Are there translated copies are available for use by the employees? Does the upper management group possess individual copies of the program?

27. Initiate the program. Is there a plan to review and critically evaluate the effectiveness of the program at least one time per month for the first six months? When deficiencies are identified, are there plans prepared to make the necessary changes/modifications while ensuring compliance with the OSHA standard or agency regulation?
28. Is there a developed safety and resource control audit assessment procedure and instrument? Is there a plan for auditing the programs on a periodic basis? Is there an audit schedule?

As can be seen, one size does not fit all safety and resource control programs. It is usually the responsibility of the safety and resource control professional to identify the requirements and needs and develop specific customized programs to achieve compliance or specific objectives. Safety and resource control professionals can, and should, learn from the past, from others' successes, and from all sources. However, all programs must be customized to fit the needs and resources of the individual situation, organization, workforce, and operation.

ENDNOTE

1. *Managing Employee Safety and Health Manual*, Tel-A-Train, Inc. (1991).

CHAPTER 6

CULTURAL SHIFT AND PSYCHOLOGICAL NEXUS

> One reason why men and women lose their heads so often is that they use them so little! It is the same with everything. If we have anything that is valuable, it must be put to some sort of use. If a man's muscles are neglected, he soon has none, or rather none worth mentioning. The more the mind is used the more flexible it becomes, and the more it takes upon itself new interests.
>
> —George Matthew Adams

> The empires of the future are the empires of the mind.
>
> —Winston Churchhill

In a utopian safety world, employees would possess a desire to work safely, would never take chances, would work with perfect equipment, would always be adequately trained, would work in a perfect environment and would never have "accidents." However, in our "real world," the myriad of imperfections has resulted, in essence, for the need for the safety and resource control professional.

Fundamentally, who is responsible for safety and resource control in the organization? Is it the safety and resource control professional? Is it the chief executive officer (CEO) and upper management? Is it the first line supervisor or team leader? Or is it the employees? Although the safety and resource professional may receive the title and shepherd the overall efforts, to be truly effective, every level and function within the organization must accept and make safety and resource control part and parcel of their job responsibilities. Everyone is responsible for safety and resource control.

This sounds great in theory, however can safety and resource control professionals make this work in the real world? Employees possess "baggage" from a number of outside sources, including home stress, relationship stress, financial stress, and other sources, which they bring with them to the workplace everyday. The workplace adds additional pressures and stressors, such as fear of losing one's job, downsizing, rightsizing, job responsibility modifications, a poor superior, harassment, and other stressors. Each of these stressors experienced by individuals on any given day is provided a priority that garners their attention for a period of time. Working safely may or may not be in the top category of the individual's priority list every minute of the working day. For example, if an employee has a sick child at home, that employee's physical body may be making widgets, but his mind is on the well-being of his child in the day care center.

Although safety and resource control professionals would prefer to be able to control every factors directly related to the causes of accidents, the thought patterns of employees is beyond our reach. However, safety and resource control professionals can effectively manage the behaviors that may result from the stressors being placed on the individual. In essence, acceptable behaviors are rewarded and unacceptable behaviors are chastized in a progressive manner, to permit the individual to transform the unacceptable behaviors into acceptable behaviors. In the event that an employee fails to transform these unacceptable behaviors into acceptable behaviors, employment is severed. In human resource terms, this is called a *progressive disciplinary system.*

The issuance of discipline to modify employee and management team member behavior is often one of the most difficult tasks performed by the safety and resource control professionals or other designated people within the organization. Modifying behaviors through the use of positive or negative reinforcement is confrontational by its very nature and has recently become a legal minefield. In most companies and organizations, the "things" that employees and management team members do at work, from showing up to the job to confrontations with other employees, must be monitored and addressed in a manner through which the company or organization can maintain control of the workplace.

Traditionally, modification of unacceptable behavior in employees was through negative reinforcement (i.e., progressive disciplinary system), whereby the employee would be provided a series of increasingly severe penalties for nonconforming behavior until the ultimate step of discharge or termination was earned. Recent research has found that more efficient modification of unacceptable employee behavior can be achieved through the use of positive reinforcement, rather than negative reinforcement. However, many laws, collective bargaining agreements, management theories, and other requirements mandate the use of negative reinforcement. Many companies are now incorporating the benefits of using both positive reinforcement through coaching and counseling along with the negative reinforcement of a progressive disciplinary system to achieve an acceptable level of behavior and performance in their workforce.

Safety and resource control professionals should be knowledgeable in the various techniques that can be used to modify unwanted or improper employee and management team member behavior in order to maximize the efficiency of the operations. Safety and resource control professionals should be aware of the numerous legal pitfalls that can be encountered during this behavior modification process.

It is vitally important that safety and resource control professionals identify the specific status of their company or organization in order to develop and use the most effective methods of proper positive and negative reenforcement for their workforce. Some companies and organizations use the positive reenforcement technique of coaching while other companies and organizations use only the counseling technique. Some companies do not use any type of positive reenforcement primarily because of collective bargaining agreements (i.e., union contracts) or other requirements, whereas other companies and organizations follow extensive positive reenforcement procedures before initiating any disciplinary action. To assist you in identifying your current status and identifying your individual perimeters, the following series of questions can be used to analyze and ascertain individual status:

- Does your company or organization have a union contract or collective bargaining agreement?
- Is your disciplinary policy set forth in a handbook or written policy?
- Is your current disciplinary policy set forth in written form?
- Are you using any type of coaching?
- Are you using any type of counseling for employees or management team members?
- What type of disciplinary policy have you been using in your facility?
- What do you your employees expect under your current disciplinary policy?
- Has your disciplinary policy, as written, been enforced?
- What is the process used to evaluate your disciplinary policy?
- How are employees monitored with regard to their performance?
- How are employees monitored with regard to their absenteeism?
- What is your past practice with regard to employee discipline?
- Have you terminated any employees under your current disciplinary policy?
- What are the steps in your current disciplinary policy (i.e., verbal, written, suspension, termination)?
- Do you suspend employees while investigating?
- Do you use any type of positive reenforcement for employees?

Discipline: Usually a formalized progression of disciplinary actions intended to identify and correct unwanted behaviors. Under most progressive disciplinary policies, the severity of the disciplinary action increases with recurrence of unwanted behaviors, with the final level being involuntary termination from employment. For example, the progressive disciplinary system in some companies includes a verbal warning, followed by a written warning in the employees personnel file. If the unwanted behavior has not changed, the next step is a 3-day suspension without pay. The final step is involuntary termination from employment.

In some companies or organizations, the first level of formal identification of unwanted behavior and positive reenforcement used is

EXAMPLE: Rule Violations

Extreme Rule Violations

Disciplinary Action

First Offense: Termination

Examples include

- Involved in a fight in the plant.
- Failing to follow the Control of Hazardous Energy Policy (lockout/tagout).
- Intentional destruction of another team members' or company's property.
- Possession or use of any intoxicating beverage or controlled substance on the company's premises (including the parking area).
- Stealing another team members' or company's property.
- Immoral conduct on company property.
- Falsification of information on employment application.
- Conviction for any criminal act.
- Employee falsely reporting off sick or falsely reporting a work-related injury.
- Walking off the job (considered de facto resignation).
- Knowingly violating OSHA standards, EPA regulations, USDA regulations, or other regulations or laws.
- Reckless conduct (i.e., throwing a knife, failure to follow physicians' instructions).

Serious Rule Violations

Disciplinary Action

First Offense: Three days off without pay
Second Offense: Termination

Examples include

- Smoking in unauthorized locations on plant property.
- Insubordination or using abusive language to team leader or management team member.
- Unexcused absences.
- Habitual lateness.
- Repeated absenteeism.
- Failure to use the required personal protective safety equipment as required for your job (e.g., safety glasses, shoes, hats, ear plugs).
- Failure to carry out reasonable instructions.
- Gambling during working hours.
- Horseplay if it could result in personal injury or property destruction.
- Coersion (in any form) of another employee for the purpose of restricting production.
- Failure to report injuries to self or others that occur on the job.
- Sleeping on the job.
- Jeopardizing property through careless action or poor work.

Less Serious Rule Violations

Disciplinary Action

First Offense: Written Warning
Second Offense: Three days off without pay
Third Offense: Termination

Examples include

- Entering another person's time card in the time clock.
- Poor production performance.
- Production of scrap or waste product because of negligence.
- Not observing correct times for lunch breaks and wash-up periods.

- Running in the plant.
- Reckless driving in the parking lot.
- Using unauthorized exits.
- Tampering with or defacing authorized bulletin boards, fire extinguishers, alarms, and other devices.
- Violation of good housekeeping and sanitation rules.
- Personal hygiene.
- Loafing on the job.

that of coaching. Coaching is a positive reenforcement tool that provides the employee with an opportunity to review the unwanted behavior and positive reenforcement as to the acceptable behavior.

Some safety and resource control professionals use the term "coaching" to describe any discussion intended to improve employee performance. However, in this module, "coaching" refers to a formal, documented and very structured discussion aimed at getting the employee to contract with you to improve a specific behavior. In many companies, coaching is performed by the supervisor or team leader, while in other companies the coaching sessions are performed by the safety or human resource professionals. In either case, the safety and resource control professional must possess a firm grasp on this important modification technique.

In many respects, a formal coaching session is similar to a intervention by the supervisor or team leader on the work area. For example, both techniques involve specific steps that need to be followed in order for the process to work effectively. Just as in a work floor intervention by the supervisor or team leader, safety and resource control professionals want the employee to do most of the talking. The safety and resource control professional's basic role is to prompt the employee through the process of asking questions.

Perhaps the most significant difference between intervention and a coaching session is that a coaching session is more of a problem solving session. The focus is primarily on the employee's willingness to change the unaccepted behavior, rather than his or her ability to exercise better judgment.

The first step in ensuring that a coaching session is as effective as possible, the safety and resource control professional needs adequate preparation. Coaching sessions should be held in a office area or other private location and should not be conducted on the work floor if at all possible.

Even if the coaching sessions are not considered disciplinary actions under your current disciplinary policies, the safety and resource control professional should treat these coaching sessions in a disciplinary manner. Coaching should be a formal, planned session and should always be documented!

It is generally not a good idea to begin a coaching session immediately after observing an employee working improperly, especially if you are anger or frustrated by the employee's behavior or comments that may have been said to you during the intervention. It is better to wait until you and the employee are both calm. In addition, there are a few things you need to do before beginning the coaching session:

- The safety and resource control professional should choose a time when it is possible to allow the coaching session to run as long as necessary. The time allotted will vary depending on the coaching skills and the employee's frame of mind.
- The safety and resource control professional should take whatever steps necessary to avoid being interrupted by telephone calls, visitors, and other responsibilities. Once the coaching session has begun, it should continue until the ultimate goal is accomplished.
- The safety and resource control professional should have specific details about the particular unwanted behavior that will be addressed during the coaching session. The safety and resource control professional should review the necessary documents and be able to refer to specific dates, times, and comments.
- The safety and resource control professional should be crystal clear about all the possible consequences if the employee continues the unwanted behavior, as well as how the acceptable procedure or behavior should be conducted.
- The safety and resource control professional should consider before the coaching session what needs to happen at the end of the coaching session for you and the employee to deem this session successful. Determine the acceptable alternatives that may be possible and the time frame in which you expect the employee's performance to improve.

EXAMPLE: Coaching Report

Name of Employee: ______________ Date: ______________

Coach's Name: __________________

Reason for Coaching Session: ________________________________

____

____

Location of Coaching Session: _______________________________

1. What was the unacceptable behavior? ____________________

   ____

2. Did the employee acknowledge the unacceptable behavior?
   Yes No
3. What are the employee's reasons for the unacceptable behavior? ____

   ____

   ____

4. Should any other department be notified regarding these reasons? Yes No

   If yes, please identify? __________________________________
5. Did you problem solve with the employee? Yes No
6. What solutions were identified during this session? _________

   ____

7. Did the employee agree with the solution(s)? Yes No

   If No, please explain: ___________________________________
8. Did the employee identify changes to be made in his/her behavior? Yes No
9. Did the employee agree to modify his/her behavior?
   Yes No

tive bargaining agreements (i.e., union contracts) normally have their progressive disciplinary action procedures codified in writing and included within the written agreement. Some companies also have progressive disciplinary systems set forth in accordance with other contractual relationships or for other reasons. However, virtually all disciplinary policy and procedures have the following characteristics:

- Virtually all disciplinary policies are in writing and are posted in some manner for employees review. This may include inclusion in employee handbook and union contract.
- Virtually all employees have been been provided some type of training in the required steps of a progressive disciplinary procedure in some manner.
- In virtually all disciplinary policies and procedures, the following statement is provided to warn employees and provide wide management discretion. Failure to take the necessary and appropriate corrective action(s) may result in additional progressive disciplinary action up to and including discharge.
- The first step of a progressive disciplinary system is normally a minor verbal warning or "slap-on-the-hand" warning.
- Subsequent steps within the disciplinary procedures normally include but are not limited to, a written warning in the employees personal file, and suspension for a number of days.
- Many companies also provide for a procedural suspension without pay pending any type of investigation as to unwanted workplace activity or actions.
- The final stage in most disciplinary procedures is the involuntary termination of the employee or management team member from their employment. This is known as the "workplace death penalty," given the fact that the employee has had a number of levels with positive and negative reenforcement before reaching this ultimate level.
- In some companies, the employee is permitted to voluntarily terminate employment or provided other routes through which to avoid the involuntary termination. This is often used for management team members.
- In virtually every phase of the disciplinary action system, documentation is essential to ensure complete and total accuracy.

> Personal and human resource managers should remember that you are affecting the livelihood of employees and management team members, and the extent of your actions goes beyond the plant gates affecting the employee's family, reputation, and so forth. All documentation should be completely accurate at all times.

Safety and resource control professionals should be aware that formal progressive disciplinary actions are negative reenforcement for employees and management team members. This procedure is also governed by very strict federal, state, and often local laws, and the specific legal ramifications must be considered throughout the disciplinary procedure. Thus, complete and accurate documentation is essential to prepare for challenge of the disciplinary action in internal and governmental courts and tribunals.

Although you may be uncomfortable issuing this type of negative reenforcement, it is essential for the effective performance of the employee as well as the success of the company. Remember, every phase of the disciplinary action must be documented in detail, as most situations are viewed with the luxury of hindsight by a court of law.

The critical part of any approach to improving employees performance is to determine which behaviors need to be changed and which ones need to be reenforced. This means that you have to have a reliable method of routinely monitoring your employees performance in the area of attendance, work performance, and safety. Especially in the area of work performance, the only real effective way to do this is for your supervisors or team leaders to periodically observe the employees work habits.

A simple, but very effective way to monitor employees work performance, safety performance, and other requirements on the shop floor is to specifically set aside time in an organized program for your supervisors or team leaders to concentrate their full attention exclusively to monitoring their employees at work. This type of program, often called *behavioral focused observations,* does not take a lot of time for safety and resource control professionals or designees (e.g., supervisors) to perform. In fact, by routinely concentrating only on employee work performance or safety performance for just a minute or two, the supervisor or team leader can quickly get a very good idea

EXAMPLE: Disciplinary Policy

Absenteeism

Any team member who is unable to come to work should notify ________________ at ( ) _____-___________ as promptly as possible, but in no case more than 4 hours after the beginning of his/her shift. Those team members calling four hours after their shift has begun will be marked unexcused for that day. Thereafter, the team member is to report to the human resource professional's office or other designated office after each 2-week period as long as he/she is off work.

Excused absences are granted for sickness if reported properly, and for personal reasons, if requested and approved the day before by the employee's team leader.

Any team member absent for more than three days must have a signed excuse from the doctor, turning it into the human resource professional before starting their work day. Any team member not having a doctor's slip will have an unexcused absence.

Unexcused absences will not be tolerated.

Disciplinary procedure in cases of unexcused absences:

First Offense: Three days off without pay (first offense in a 1-year period).

Second Offense: Termination (second offense in a 1-year period).

All team members missing six times or more in 6 months (except for vacation, funeral, or disability days) will have a copy of their attendance sheet sent to the plant manager by their team leader.

Disciplinary procedure in cases of excessive absenteeism:

1st Offense: Written warning (sixth absence in six-month period and/or a continuous poor record).

2nd Offense: Three-day suspension without pay (7th absence in 6-month period or continuation of absenteeism).

3rd Offense: Termination (8th absence in 6 months or a continuation of absenteeism, i.e., several warnings in successive years or additional absenteeism).

Any team member calling in to report that he or she is sick who is later found not to have been sick will be discharged.

Any team member having excessive absence because of sickness will be required to have a doctor's examination and then will furnish his team leader or supervisor with a complete doctor's report. If the team member is eligible for a disability pension, this possibility will be investigated.

on how much attention he or she needs to provide to the individual employee.

To ensure that the observation program is effective, there are a few basic guidelines for safety and resource control professionals to follow:

- Employees may be uncomfortable if the safety and resource control professional or supervisor simply begins watching them without first explaining their intentions. They maybe concerned that the supervisor or safety and resource control professional are trying to catch them doing something wrong, and if they do, they will be disciplined. The supervisor or safety and resource control professional needs to make it clear, at the very outset, that their objective is to identify behaviors that need to be improved, before they can lead to an accident or work performance problems. Once the employee realizes that the supervisors or safety and resource control professional is interested only to benefit them, it will be a great deal easier for them to get the support they need from the employee.
- Supervisors or safety and resource control professionals need to observe employees work habits, safety procedures, and so forth as frequently as possible. Ideally, the supervisor or team leader should interact, communicate, and observe every one of their employees on a daily basis.
- Supervisors or safety and resource control professionals not only need to identify unwanted behaviors but should also praise appropriate behaviors. The most powerful tool that a supervisor or safety and resource control professional can use to motivate

CHAPTER 7

MY PROBLEM EMPLOYEES

> I tell you, sir, the only safeguard of order and discipline in the modern world is a standardized worker with interchangeable parts. That would solve the entire problem of management.
>
> —Jean Giraudoux

> People work for people, not for companies. A worker's regard for his supervisor will affect his opinion of his employer. Production is related to attitude, so much so that an organization which disregards this human equation will not achieve as much as it could achieve.
>
> —Gerard R. Griffin

Safety and resource control professionals often use the axiom that they spend 90% of their time with 10% of their employees and 10% of their time with 90% of their employees. Disgruntled employees, as with constant offenders, can require a substantial amount of your time due to the sheer number of sessions in an attempt to modify their unwanted behaviors. Disgruntled employees, unlike constant offenders, often possess an exceptional attendance record and work performance record and could be exceptional employees if they did not constantly complain about virtually everything.

Disgruntled employees or "constant complainers" have underlying reasons for their inappropriate behavior, which can include the following:

- Loss of control—the disgruntled employee may be in a job in which he or she has little control.
- The disgruntled employee needs to voice his or her opinion.
- The disgruntled employee is getting no response or feedback with regard to his or her complaints.
- The disgruntled employee may be serving as the unofficial leader for the work area.
- The individual may be having personal difficulties outside of work that are being vented through their actions on the job.

Safety and resource control professionals should attempt to identify the underlying reason for which the employee has become disgruntled or is constantly complaining. Often, involvement with company activities, such as the safety committee, can offer the employee a method through which to voice any concerns in the appropriate manner. Safety and resource control professionals should not completely discredit the complaints that come from a constant complainer, given the fact that the "allegedly" disgruntled employee may be the unofficial leader of a group of employees who are experiencing problems in their work area. Safety and resource control professionals should strive to turn what appears to be a negative behavior into a positive behavior for the employee, through appropriate means within their organization or company.

Given the increased incidents of workplace violence in America, it is important for safety and resource control professionals to adopt appropriate protection to safeguard the human assets of the company, especially when issuing coaching, counseling, or discipline. The primary areas of potential risk to persons and property are in the final stages of the disciplinary action policy or when individuals return to the facility for some reason, such as a workers' compensation check, after they have been terminated from employment.

Safety and resource control professionals must anticipate the potential problems that could resolve and prepare to address these potential problems. These can include:

- Angry employee venting physical aggression toward personnel and human resource manager.

- Terminated employee causing property damage while remaining in the plant.
- Terminated employee causing damage to automobiles or property in the parking lot.
- Angry employee causing physical violence to other employees.
- Angry employee returning to the plant with weapons.
- Angry employee sabotaging operations or facilities.

In many companies, the use of procedural suspensions pending investigations with written notice of termination to the employee has minimized the potential exposure for many personnel, safety, security, and human resource managers. Not only does this method provide the manager with the opportunity to investigate thoroughly the allegations against the employee, but it also prevents the possibility of physical confrontation with the employee on company property. Example, the following is an example of the use of procedural suspensions:

- Employee performs a terminable offense (e.g., violation of work rules)
- Employee is brought to the personnel and human resource office (or other designated location) for disciplinary action.
- Safety and resource control professional and/or others acquire a written statement from the employee as to the reasons for the actions.
- Employee is placed on procedural suspension and escorted from the facility.
- Safety and resource control professional thoroughly investigates the allegations and finds that they substantiate a terminable offense.
- Safety and resource control professional notifies employee of procedural suspension by written correspondence (certified mail).
- Final assessment and determination. Employee terminated. Employee notified of termination by certified mail.
- Terminated employee is required to report to the facility in a specified time and all personal items are removed from the locker,

DISCIPLINARY CHECKLIST

- Have you reviewed the documentation to ensure complete accuracy with regard to the disciplinary action?
- Have you followed all company policies and procedures?
- Where is the disciplinary action to take place?
- How are you going to document the disciplinary session?
- Have all participating parties been notified (e.g., business agent, witnesses, security)?
- Has security or management been notified to escort the employee from the facility after the disciplinary session?
- Does the employee possess violent tendencies?
- Can the employee gain access to company property or personal items (like your car) after the disciplinary session?
- Has a procedure been instituted to prevent the employee from reentering the facility after the disciplinary session?
- Was the employee given an opportunity to explain his/her actions? Was the employee permitted to "vent" his/her emotions?
- Were all exiting documents (e.g., insurance) completed? Are these documents forwarded to the employee's home residence?
- Is you file complete and prepared for litigation?
- Other company requirements: ______________________________

____

____

desk, etc., and the employee is escorted from the facility by security or other management team member.

- Terminated employee is not permitted into the work areas, nor is he permitted to any contact with any of the parties involved in the termination.

Safeguarding yourself, other employees, and company property both during and after disciplinary actions is essential. Safety and resource control professionals should take every effort through which to anticipate any type of violent reaction by the employee before, during, or after any type of disciplinary action. The safety and resource control professional should attempt to defuse and type of violent actions by the employee and should take every possible precaution to avoid any future encounters with the employee after termination.

Generally, employees always know what is expected of them, as well as what is right and wrong, first, because they are mature persons who have their own common sense, and second, because they have been trained to perform the requirements of their respective jobs. In the event that an employee or team member engages in errant behavior, it is an inherent management right, and it is the company's policy, that such team member be dealt with as the individual that he/she is. Basic fairness requires that, before any disciplinary action results in lost time, the team member be given is given a Notice of Procedural Suspension Pending Investigation.

The delivery of such a notice is not intended as the imposition of discipline, which can only be effected as hereinafter provided. Such notice should be in writing, but if service of the notice is not practicable, the notice will be sent to the team member immediately. On the face of the written notice is a provision requiring that the team member return to the company's offices or other designated office for the purpose of giving his version of the subject incident and to be asked if he wants management to interview any other persons on his behalf, as part of management's investigation of the subject incident. The company's management team will endeavor to complete an objective investigation with special emphasis on the subject of disparate treatment. That is, if discipline were imposed on the subject team member, would such treatment be unfair, as compared with how management has treated other employees who are similarly situated? The company's management team with then prepare a written report of its objective investigation. Such report will contain findings of fact and whether any discipline is recommended. Such report will conclude with an analysis of any alleged disparate treatment. If the management team concludes after such investigation that no discipline is warranted, the team member will be reinstated with full back pay and no loss of benefits or status.

Because the circumstances surrounding each case of errant behavior are rarely identical to any other case, no set disciplinary action will be announced in advance. Each case is considered on its individual merits.

Safety and resource control professionals are often faced with the opportunity of exhaustive and extended litigation after the termination of an employee or management team member. As described throughout this text, documentation is essential in cases where an employee believes he/she has been wrongfully terminated or unjustly treated by the employer. The safety and resource professional, as the representative of the company, is often called upon to testify in various types of legal forums in an attempt by the court to evaluate the justness of the termination. The safety and resource control professional must be prepared for other allegations, such as filing a OSHA complaint or an Equal Employment Opportunity Commission (EEOC) charge of discrimination, to appear after the termination from employment. Preparation and documentation are the keys to the successful defense of such allegations in the future.

What can a safety and resource control professional expect after a termination of employment by an employee? Although every case should be analyzed on a case-by-case basis, the following are some of the potential areas of litigation that may ensue:

- Unemployment insurance claim.
- Workers' compensation claim.
- Charge of discrimination for reporting an OSHA complaint.
- Charge of discrimination for filing a workers' compensation claim.
- Charge of discrimination for sexual harassment.
- Charge of discrimination based on age.
- Charge of discrimination based on race.
- Charge of discrimination based on religion.
- Charge of discrimination based on national origin or color.
- Wrongful discharge action based on an exception to the at will doctrine.
- Charge of breach of contract.
- Union grievances.

The importance of proper documentation in the defense of any of the above allegations or claims cannot be overemphasized. Given that the safety and resource control professional is often the focal point of the disciplinary action, it is not unusual for the safety and resource control professional to be the party named in any of these actions or disposed and required to bring all documentation to deposition. The agencies and courts evaluating these cases have the luxury of hindsight through which to analyze and evaluate your decision making process. Safety and resource control professionals should be knowledgeable in this area and prepare their documents in a defensible manner in anticipation of this type of litigation. If litigation or claims does not happen, it is easy for the safety and resource control professional to dispose of these documents in the normal course of business. However, if these documents are needed, it is essential that the safety and resource control professional possess these documents, these documents should be in appropriate order defended against the claim.

One of the job functions of most safety and resource control professionals is to mold and modify the behaviors of employees and management team members in your company or organization. Through the use of positive reinforcement techniques such as coaching and counseling, combined with the negative reinforcement provided by disciplinary action, this can usually be accomplished. If the safety and resource control professional is unable to modify unwanted behaviors in the employee or management team member and the "workplace death penalty" results, the company or organization loses thousands of dollars that have been invested in this "human asset." The impact on the individual, the family, the morale of co-workers, and many other types of losses are also incurred.

Safety and resource control professionals are responsible for protecting and managing the human assets of the company or organization. Through the use of coaching, counseling, and disciplinary action, appropriate modifications can be made to unwanted behaviors to achieve a satisfied and productive employee or management team member. The results of your efforts can be enormous for your employees, your management team members, and your company or organization.

CHAPTER 8

COMPLIANCE AS A FOUNDATION

> We are not free to use today, or to promise tomorrow, because we are already mortgaged to yesterday.
>
> —Ralph Waldo Emerson

> Study the past if you would divine the future.
>
> —Confucius

The ultimate goal for every safety and resource control professional is to safeguard employees from harm in the workplace. A secondary goal, although vitally important, is the achievement and maintenance of compliance with the Occupational Safety and Health Administration (OSHA) standards and requirements. In order to reach these important goals, a comprehensive management approach needs to be developed to manage the safety and resource control function and an extensive, all inclusive strategy through that to direct and control the completion of the required tasks in order to achieve compliance with the OSHA standards and regulations.

A management philosophy should be incorporated to serve as the foundation of the safety and resource control function and to provide the necessary style through which to manage the safety and resource control function. The selection of that management philosophy and style is an individual decision based on the background and personality of the safety and resource control professional, type of industry, employee population, and numerous other factors. The key to achieving compliance with the OSHA standards and regulations is to the safety and resource control function, rather than permitting the safety and health issues or problems to dictate to ones organization.

The management principles used by management team members in daily supervision of production, quality control, or any other operation are the same when managing safety in the workplace. In production you plan, organize, direct, and control your operation to produce a product while in safety you plan, organize, direct, and control[1] the safety and health of the employees in the workplace.

PLAN OF ACTION

In developing the appropriate mechanisms to manage OSHA compliance in the workplace, a written plan of action is one of the initial steps. A written plan of action, not unlike a battle plan in military terms, sets forth the objective of each activity, delineates the activity into smaller manageable elements, names the responsible parties for each element of the activity, and provides target dates at which time the responsible party will be held accountable for the achievement of the particular element or activity. In order to manage this planning phase, safety and resource control professionals can use a planning document that permits the safety and resource control professional to evaluate progress toward the objective on a daily basis and to also hold the appropriate responsible party accountable for the achievement of the particular element or activity. This type of planning document can be computerized or simply completed in written form.

In developing a plan of action, all levels of the management team need to be involved in developing priorities and scheduling the plan of action. This team involvement assists the management team members to buy in to the overall safety and resource control efforts of the organization. It also permits input regarding potential obstacles that could be encountered and the development of a realistic targeted time schedule given the OSHA time requirements and worksite pressures. Ranking of the various mandated safety and health programs and other safety and loss prevention programs that are not required by

OSHA should be given careful attention so that the appropriate programs affording employees the maximum protection and meeting the OSHA target dates are given first preference.

Management team members should be advised that team members will be held accountable for the successful and timely completion of their assigned tasks and duties as set forth under the plan of action. With the required management commitment to the safety and resource control goals and objectives by top management and management team member's "buy in" during the development of the plan of action, management team members should be well aware of their specific duties and responsibilities within the framework of the overall safety and resource control effort of the organization. If necessary, appropriate positive or negative reinforcement can be used to achieve this purpose, in addition to appropriate disciplinary action.

Safety and resource control professionals should be cautious when developing written safety and health programs to meet compliance requirements. The methods used in the development and documentation of OSHA compliance programs are a direct reflection on the safety and resource control professional and the safety and health efforts of the organization. Safety and resource control professionals should always develop written safety and health compliance programs in a professional manner. Safety and resource control professionals should be aware that when compliance officers are evaluating the written compliance programs, this could possibly set the tone for the entire inspection or investigation.

Compliance programs should also be developed in a defensive manner. Every element of the OSHA standard should be addressed in written form and all training and education requirements should be documented. In the event of a work-related accident or incident that places the written safety and health programs "on trial," a program will be placed under a microscope and every detail scrutinized from every angle. Proper preparation, evaluation, and scrutiny when developing a written compliance program can avoid substantial embarrassment, cost, and other liabilities in the future.

In the area of required education and training elements mandated under a particular OSHA standard, documentation is vital. Safety and resource control professionals should closely evaluate the compliance program to ensure that all required training and education mandated under the standard is being completed in a timely manner.

In addition, the training documentation should be clear, to confirm, beyond a shadow of a doubt, that a particular employee did attend the required training. This documentation should show that an individual employee not only attended during the training and education session, but understood the information provided. To show understanding and an adequate level of competency, a written examination may be helpful. If a required training and education element is not documented, there is no type of proof to substantiate that employees under went a certain type of training.

In the area of training and education elements with a compliance program, safety and resource control professionals are reminded of the educational maxim, "tell them, show them, and tell them again." Safety and resource control training should be conducted, where feasible, in an atmosphere conducive to learning and at a time when employees are mentally alert. The individuals performing the training should be competent and enthusiastic. Hands-on training has been found to be the best method that provides the greatest understanding and that employees retain the information better. Audiovisual aids are an exceptional method of increasing the retention level but safety and resource control professionals should not rely solely on the audiovisual aid (i.e., especially videotape) for the total training experience.

Although safety and resource control is a serious matter, training does not have to be a sober and boring endeavor that employees are required to endure on a periodic basis. Safety and resource control professionals should strive to make training an experience that attendees will remember. There is no rule that safety and resource control training cannot be mentally stimulating or even fun. Remember, the information provided in a safety and resource control training session may mean the difference between an employee who goes home and one who does not, at the end of the day—do everything possible to ensure that the employee is provided, understands, and retains the information from the training session.

Purchasing the appropriate personal protective equipment for the particular circumstance in order to achieve the objectives of the OSHA standard is vitally important. Safety and resource control professionals should take an active involvement in the selection, purchase, monitoring, inspection, and replacement of personal protective equipment. Although cost is always a factor, the safety parameters, comfort levels, approval or certifications, and other factors need to be scrutinized to ensure that the personal protective equipment meets or exceeds the requirements mandated under the OSHA standard and, that the personal protective equipment is of a type and quality that will not cause employees difficulties in everyday use. Many safety and resource control professionals provide the initial evaluation and selection of broad types of personal protective equipment and permit the individuals who will be required to wear the personal protective equipment to make the final selection. This type of employee and management team member involvement in the selection process often leads to greater participation in the program.

If a safety and resource control professional is unsure as to the requirements of a particular standard, it is imperative that he/she acquire a definite answer or clarification. OSHA often provides clarification of a particular issue or problem without caller identification or the call may be transferred to the state education and training division. In many state plan states and in federal states (usually through a state agency), a separate section of OSHA has been established to assist employers in achieving compliance. Upon request, the education and training section can assist employers with a wide variety of compliance issues ranging from program development to the acquisition of pertinent information at no cost. Safety and resource control professionals should be aware that this section of OSHA does possess the ability to issue citations, but normally does not issue citations except in situations involving imminent harm or if the employers failure to follow prescribed advise.

Finally, safety and resource control professionals should acquire a strategy to effectively manage a number of compliance programs simultaneously. The use of a safety and resource control audit can be an effective tool in identifying deficiencies within a compliance program and permit immediate correction of the deficiency. There are various types of safety and health, security, loss prevention, and related audit instruments, but all audits include the basic elements for identification of the required elements of a compliance program: track the current level of performance, identify deficiencies, and identify potential corrective actions. A safety and health audit mechanism can provide numerical scoring, letter or grade scoring, or another method of scoring, so that the management team can ascertain their current level of performance and identify areas in need of improvement.

In achieving compliance with an OSHA standard, it is imperative to check and contain that every area has been looked into and is in compliance. To this end, below is a basic evaluation instrument to assist the safety and resource control professional in addressing potential areas that may have been overlooked:

1. Is the employer covered under the OSH Act?
2. Has the facility or worksite been evaluated to ascertain which specific OSHA standards are applicable?
3. Has the management group been educated as to the requirements of the OSH Act and standards and acquired the necessary support and funding for the programs?
4. Is there a copy of the OSHA Standards (29 CFR 1910 et seq. and other standards) at the worksite?
5. Is the Federal Register or other appropriate sources for new standards or emergency standards being reviewed to see whether they are applicable to the worksite?
6. For new programs or new standards promulgated by OSHA, is the OSHA standard applicable to the workplace, situation, or industry?
7. Is there an OSHA Guideline for particular situations or hazards in the workplace?
8. If there is no applicable OSHA standard, will the situation qualify under the General Duty Clause as an unsafe or unhealthful situation or hazard?
9. If there is no applicable OSHA standard and the situation is deemed to qualify under the general duty clause, have National Institute of Occupational Safety and Health (NIOSH) publications, American National Standards Institute (ANSI) standards, or other applicable journals and texts been reviewed to acquire guidance?
10. If there is no applicable OSHA standard, or there is an OSHA standard that conflicts with other government agency regulations, there are other options: (1) contact the regional OSHA office and request consultation assistance, or (2) employ an outside consultant possessing the specific expertise to assist, or (3) pursue a variance action?

11. If there is an applicable OSHA standard, has each and every word of the standard been read so as to understand completely the requirement of the standard?
12. Is there a written program to ensure compliance with all requirements of the OSHA standard? Remember, if the program is not in writing, then there is no evidence to prove the existence of the program during an OSHA inspection or in litigation. Is the original of this written program in a secure location?
13. Is the program written in a defensive manner? Can the written program be scrutinized by OSHA or a court of law without identifying flaws in the program? Is the program written in a neutral and nondiscriminatory language?
14. Is the documentation of every purchase, equipment modification included in the program, in order to ensure compliance with the OSHA standard? Is this documentation in the original copy of the written program or in a secure location?
15. Does the written program possess the purpose of this program? Is a copy of the applicable OSHA standard available for easy reference? Have the responsibilities been delineated, and are they specific for each level of the management team and each position within the levels of management?
16. Has the OSHA standards and other applicable information been closely scrutinized and evaluated? Has each and every requirement in the standard been achieved or exceeded? Have any steps or elements required in the OSHA standard been omitted? If the standard is vague, make sure the program is clear, concise, and to the point.
17. Has all training been documented as required under the OSHA standard? (Remember, the OSHA standard is only a minimum requirement. Every program can be better than, but not less than, the OSHA standard requirements.)
18. Is there detailed documentation regarding each and every phase of the training? Is this documentation in the written program? Is documentation provided in the written program to prove the use of audiovisual aids, the instructor's qualifications, and other pertinent information?

19. Is there a schedule of the classroom and hands-on training sessions in the written program?
20. Have all employees who have completed the required training signed a document showing the exact training completed (in detail), the instructor's name, and the date of the training? Have auxiliary aids and other accommodations for individuals with disabilities been provided for in the training programs? (For employees who cannot read/write, a thumb print can be used. Videotape documentation of the training is also an acceptable method of documentation. Remember to maintain the individual tapes on file, as with other documents, for future use as evidence of this training.)
21. Is the training offered in the languages used by the employees? Are the documents and the written program interpreted into the languages spoken by the employees? Are the interpreted programs provided for use by these employees?
22. Is there a posting requirement under the applicable OSHA standard? Is the necessary poster from OSHA in the facility? Has it been posted in an appropriated location by the required date? Have posters in the language of the employees been acquired? Are they posted in an appropriated location by the required date?
23. Are there any other requirements? Labeling of containers? MSDS sheets? Does the OSHA standard require support or information from an outside vendor or agency? Have all requests to outside vendors or agencies requesting the necessary information (e.g., MSDS sheets) been documented and placed in the written program?
24. Is there a disciplinary procedure in the written program instructing employees of the potential disciplinary action for failure to follow the written program?
25. Has the written program been reviewed before publication? Does it meet and/or exceed each and every step required under the applicable OSHA standard?
26. Has the written program been reviewed by legal counsel or the upper management group before publication? Upon completion of the acquisition of necessary approvals of the written

program, have copies of the program been made for distribution to strategic locations in the facility? Are translated copies available for use by the employees? Does the upper management group possesses individual copies of the program?

27. Initiate the program. Is there a plan to review and critically evaluate the effectiveness of the program at least one time per month for the first 6 months? When deficiencies are identified, are there plans prepared to make the necessary changes and modifications, while ensuring compliance with the OSHA standard?
28. Is there a developed safety and health audit assessment procedure and instrument? Is there a plan for auditing the programs on a periodic basis? Is the audit to be scheduled?

For many years, the safety and resource control function was a secondary job function or in many cases, an afterthought. The safety and health function was managed using a "squeaky wheel" theory. That is, the only time in which the management team paid any attention to the safety and resource control function was when the wheel squeaked and this was when after an accident had already occurred. Today, with the increasing costs of work-related injuries and illnesses, the increasing compliance requirements and liability, and other increasing costs in the area of safety and health. A proactive stance should be taken to ensure that a safe and healthful environment and all resources are protected in the workplace is essential.

Often, in order for most management groups to embrace the concept of a proactive safety and resource control program, the management group must be educated as to the cost-effectiveness of such an endeavor. Safety and resource control professionals are often able to show the monetary, as well as the humanitarian, benefits of a proactive safety and resource control program through the use of a cost–benefit analysis.

Figure 8.1 (direct and indirect costs) shows the iceberg effect of the potential cost of an accident.

Figure 8.1 is often used to exemplify the actual costs of accidents and the resultant injuries and illnesses in the workplace. When most individuals think of accident costs, the first thoughts that cross their minds are the direct costs. Direct costs include the cost of maintaining a medical facility at the worksite, the medical costs and time loss

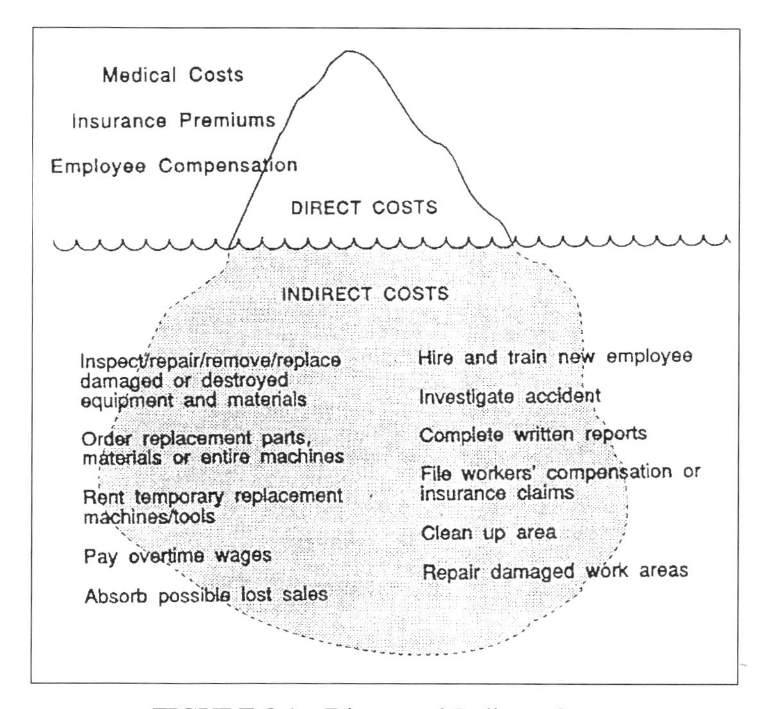

FIGURE 8.1 Direct and Indirect Costs

benefits provided under workers' compensation, and the premium costs of insurance. In most organizations, these cost figures are easily identified and able to show the management group the actual direct costs to their particular organization. These direct costs in most organizations are substantial and normally result in the use of a percentage of the profits to pay for these costs (e.g., 4% of the fiscal year 1993 profit was paid in workers' compensation benefits). When management team members actually take the time to understand the amount of money (lost profits) being spent on the direct costs of accidents, safety and resource control professionals usually obtain the immediate and intense attention of the management group.

Using the above model, the safety and resource control professional can show upper management the actual costs of work-related accidents by combining the above-water direct costs with the more elusive indirect costs shown below the water line. When the management group understands that the indirect costs of an accident can sometimes be as much as 50 times the direct costs of an accident; the safety and resource control professional is beginning to acquire the management buy-in necessary for the management commitment to the proactive safety and health effort. As can be seen from the model, indirect costs can be from equipment damage, replacement costs, quality losses, production losses, and many other areas. This model can be customized to a particular organization, using actual dollars to provide even greater impact to their presentation. This visual art works especially well with the financial number crunches in the organization.

When the management groups fully understand the cost factors involved in work-related accidents, safety and resource control professionals should also be prepared to show the dividends, in both monetary and humanitarian terms, that can be acquired through a comprehensive and systematic management approach to safety and resource control. To ensure that the management group fully understands the concepts involved in a proactive program, the management group should understand how accidents happen and how accidents can be prevented. Using the domino theory,[2] safety and resource control professionals can easily explain the causal factors leading up to a accident and the negative impact after an accident. In addition, the safety and resource control professional can explain the fact that, through the use of a proactive safety and resource control program, the causal factors that could lead to an accident can be identified and corrected before the risk factors mount, which ultimately lead to an accident. Figure 8.2 illustrates the domino effect.

Figure 8.2 shows the underlying factors that could lead to an accident or other type of loss. The safety and resource control professional should emphasize the fact that the underlying causes for workplace injuries, illnesses, and losses can be identified and corrected through the use of a proactive safety and resource control program. If the underlying factors leading to an accident are not identified and corrected, the dominoes begin to fall and once the dominos begin to fall, it is almost impossible to prevent an accident from happening. The key is to ensure that the management group realize that to prevent an accident, the underlying risk factors must be minimized or eliminated, rather than reacting after an accident has already happened.

To amplify this point, safety and resource control professionals often use the following progressional model to drive home the point

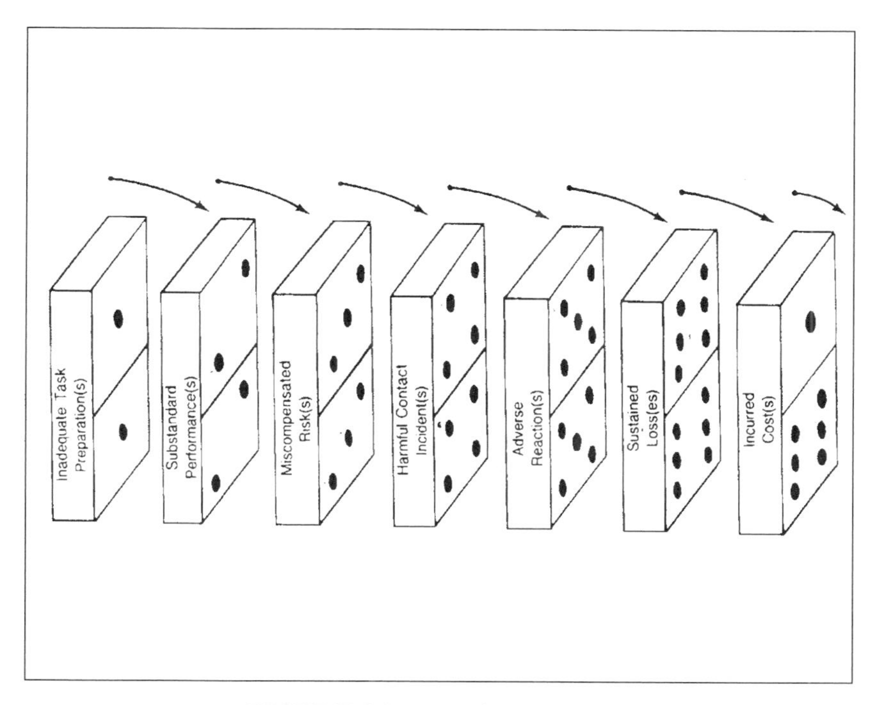

FIGURE 8.2 Domino Sequence

that near misses and other underlying factors, if not addressed, will ultimately lead to an accident. In this model, for every 300 equipment damage accidents or near misses an employer may experience, there will be 29 minor injuries. If the deficiencies and underlying risk factors are not identified and corrected, the 300 near misses will ultimately lead to one major injury or fatality. The key is to ensure complete understanding that the management team must take a proactive approach to the safety and resource control function rather than reacting when an incident or accident happens.

ENDNOTES

1. Managing Employee Safety and Health Manual, Tel-A-Train, Inc. (1991).
2. C.E. Marcum, *Modern Safety Management*, Kingsport Press, 1978, p. 29.

CHAPTER 9

ETHICAL AND PROFESSIONAL CONDUCT

> Many corporate managements and a lot of security analyst have reacted with much fuming and even more fumbling to the Texas Gulf Sulphur decision and the SEC's actions against trading on inside information.
>
> I don't know why everyone is in such a hot sweat over these two things. The one has been illegal—and damned well properly so—for over 30 years ... the initial reaction of some managements that these actions meant they could no longer talk to any publications or analysts was—is—asinine. Such sour silliness is quite the reverse of what was intended.
>
> —Malcolm Forbes

> Dishonesty, cowardice and duplicity are never impulsive.
>
> —George Knight

Where do you draw the line when it comes to virtually every area involved in the safety and resource control function? Will you destroy an important document that could be harmful to your company? Will you terminate an employee simply because your boss wants them gone? Will you fail to report a work-related injury because it will have a detrimental effect on your safety record?

Safety and resource control professionals face many tough issues in

SAMPLE: Individual Code of Conduct

1. I will assist in maintaining the integrity and competence of the safety and resource control profession.
2. I will maintain my individual competency in the safety and resource control profession.
3. I will maintain my own integrity and the integrity of the safety and resource control profession.
4. I will always tell the truth.
5. I will never provide materially false statements.
6. I will never deliberately fail to disclose material facts or documents.
7. I will never falsify any documents or statements.
8. I will not engage in illegal conduct.
9. I will not engage in acts involving moral turpitude.
10. I will not engage in conduct involving dishonesty, fraud, deceit, or misrepresentation.
11. I will not engage in any conduct that adversely affects or reflects on my fitness as a safety and resource control professional.
12. I will fulfill my duties and responsibilities to my company, organization, management team, and employees to the best of my abilities.
13. I will work with law enforcement and other authorities within the bounds of the law.
14. I will ensure the proper handling of all monies.
15. I will ensure the confidentiality of all appropriate records.
16. I will promote the safety, health, and welfare of my employees.
17. I will preserve all information provided to me in confidence.
18. I will exercise independent judgment within the bounds of my job function.
19. I will inform my employer of all potential violations of the law and encourage him/her to comply with the law. If my employer chooses to violate the law, I will take appropriate actions.
20. I will not abuse alcohol, controlled substances, prescription medication of other mind-altering substances.

21. I will not discriminate against any party.
22. I will not conceal or knowingly fail to disclose information.
23. I will not lie or provide perjured testimony or false evidence.
24. I will avoid the appearance of impropriety.
25. I will not misappropriate funds of individual or my company.
26. I will maintain fully and complete documents and records.
27. I will address situations in a timely and appropriate manner.
28. I will not attempt to improperly influence the judgement of employees, management, or others.
29. I will always strive to take the "high road" in every circumstance.
30. I will always do the best I can possibly do in every situation.

the performance of their jobs that often fall within gray areas with regards to legality, ethics, and morality. Often, the circumstances and individuals involved in the situation place the safety and resource control professional in the difficult position of making a decision when there is not a solid or acceptable alternative, that is, a lose–lose situation. Given the limited guidance in this area, safety and resource control professionals must often follow their own moral compass in making these decisions.

To avoid the off-the-cuff decision-making process, safety and resource control professional should evaluate their individual situation as well as their personal position in advance in order to determine where their personal perimeters are located and what lines the safety and resource control professional absolutely would not cross under any circumstances. This predetermination will assist the safety and resource control professional in establishing a code of conduct that can be relied on when confronted with difficult decisions.

Below is a sample Code of Conduct that can assist the safety and resource control professional in analyzing and determining the personal code of conduct or ethical rules. This sample code is only provided for guidance. Safety and resource control professionals are encouraged to add other issues or areas that may be pertinent to their individual situation and analyze each of these situations to determine individual perimeters.

Please think about the above given your current situation and strive to develop your personal code of conduct. *Ethics* is defined as the discipline that deals with what is good and bad, and with moral duty and obligations; a set of moral principles or values; the principles of conduct governing an individual or group. What are the principles you believe in? What things have you done that you are proud of? What things have you done or will do that you are not so proud of? Where are your personal perimeters as to what things you believe are right and what things are wrong?

In the end, can you say that you are proud of the work you do and that you followed the principles you believe in and have established? Right and wrong are set in our laws, but safety and resource control professionals are often faced with shades of gray.

Safety and resource control professionals who are presented with a potential conflict or ethical issue should use their personal Code of Conduct to establish the parameters. However, a systematic method through which to analyze the potential ethical issue and manage the specific problem is also necessary. The following guideline is provided to assist safety and resource control professionals in this analysis:

- Stop. Slow the situation down and ensure that you have all of the facts.
- Do not make snap judgments or decisions without all the information and evidence.
- Review all the information and evidence in detail.
- Identify possible options.
- Identify possible ramifications of each option.
- Identify short-term impacts. Identify long-terms impacts.
- Are their any options that can resolve the situation?
- Are there legal considerations?
- Is there someone you trust that you can talk to regarding the conflict?
- Who is being injured through the conflict?
- Do you have a duty to respond in any way?
- Identify your options.

- Select the appropriate option.
- Can you "live with" your decision?
- Is your decision aligned with your personal code of conduct?
- Is your decision within the bounds of your company's code of conduct?
- Is your decision within the bounds of law?
- Is the decision correct given the circumstances?
- Plan your actions. How are you going to address the situation?
- Initiate your actions.
- Reevaluate as necessary.

Remember, circumstances can change very rapidly in these types of situations. Be prepared to stop and reevaluate the situation at any time during your course of action. The decision(s) you make in these types of circumstances reflect your character and person and often have a direct impact on your career.

Safety and resource control professionals are often privy to information that is not common knowledge and is often personal in nature. Under most circumstances, the information concerns the specific company or organizations or individual issues. Safety and resource control professionals should exercise extreme caution with regard to the confidentiality of this information and how, if at all, this information is used within the job function.

For example, the safety and resource control professional is told by a supervisor that he/she has acquired a sexually transmitted disease from someone who is not their spouse during a recent business trip. In this example, the safety and resource control professional owes loyalty to the supervisor who came him or her for assistance; however, the safety and resource control professional also has a duty to the company, and potentially to the public. The implications in this type of situation can be catastrophic to many parties, including the company (e.g., workers' compensation claim, government regulations regarding contamination of the product); the spouse (e.g., confirmation of infidelity); the general public (e.g., contaminated products); as well as the individual. This type of situation places the safety and resource control professional in the very delicate position of being required to act upon the circumstance in an expedited manner because of the situation.

What would you do if your boss came into your office and asked you to hide or shred an OSHA report? A document that could be a "smoking gun" in a discrimination action? A workers' compensation claim? Do you follow orders, or is there another course of action?

As we were able to view in detail through the Iran Contra hearings, companies and organizations often attempt to cover their tracks through the elimination of documents that could be detrimental to their position. Given your capacity as safety and resource control professional, it is not outside the realm of reality for your boss or other company officials to order you to dispose of sensitive documents that you have developed during the course of your work. For example, the safety and resource control professional sends a memo to his or her boss, identifying the need for a repair on a specific machine. The boss does not act on the memo, and an employee is seriously injured. At a minimum, the boss is in "hot water" for no repairing the machine, but there is also a "smoking gun" document specifying that he or she was informed of the deficiency. The boss comes to the safety and resource control professional and asks that all copies of the memo be destroyed. Would you destroy the documents? What if your boss required you to give him or her the documents or risk termination? What would you do?

In recent years, more than a few company officials have used confidential information learned through company sources for financial gain through the stock market. The Securities and Exchange Commission (SEC) has promulgated very strict rules regarding what is commonly known as "insider trading" carrying civil as well as criminal penalties. For example, the safety and resource control professional is called by corporate and told that the company was being purchased by another company. The safety and resource control professional, knowing this will positively affect the stock price, contacts his or her broker, and buys a large quantity of the company stock. When the purchase takes place, the safety and resource control professional makes a windfall profit on the basis of this inside or confidential information.

Particularly in the areas of insurance and workers' compensation, safety and resource control professionals may be asked to neglect to file a claim form, conveniently lose the form, or hold the claim for a specified period of time. For example, the company's safety goals are established on the number of workers' compensation claims. On

December 29, an employee is injured and claim should be field. The plant manager asks you to hold the workers' compensation claim until January 1, so that the supervisors can will the bonus for achieving their safety goals. What do you do?

When litigation is likely, it is not unusual for potentially important evidence to disappear. For example, if a truck was involved in an accident and the cause appears to be the tires, the tires may be removed for closer examination. In some situations, the tires would conveniently disappear before a court order or the initiation of a legal action. What would you do if your boss told you to throw away something that you knew could be important evidence?

Sexual harassment situations can involve many ethical considerations. For example, the safety and resource control professional falls in love with an employee and the relationship goes sour. Do continued advances to rekindle the relationship constitute sexual harassment? Where are the "lines" in the area of sexual harassment? When does "no" mean "no," and when does "no" mean "maybe"?

As a general rule, "no" always means "no." The courts view the situation through the eyes of the individual being harassed. Sexual harassment is most often a male harassing a female but harassment can be a female harassing a male or a male harassing a male or a female harassing a female (known as same-sex sexual harassment). Situations involving any type of sexual harassment usually involve legal issues, policy issues, and ethical issues for all parties involved.

Although most safety and resource control professionals would not purposefully commit an illegal act, individuals can be placed in positions where the ultimate decision may result in illegal activities. This has been prevalent recently in such areas as illegal campaign contributions, falsifying DOT records, deliberately modifying OSHA records, and other violations. What would you do if your boss required you to perform an illegal activity?

Is there a difference between a thorough evaluation and investigation and a "setup"? What would you do if your boss came to you as the safety and resource control professional and told you to find a way to justifiably terminate your best friend at work? Are there ways in which an individual can be "set up" for a legal and justifiable voluntary or involuntary termination? What would you do?

Surveillance can span a broad spectrum from visual monitoring of employees on the job through concealed visual surveillance of secure areas. Legal boundaries forbid the use of surveillance in such areas as bathrooms and changing areas. Are there areas of surveillance that may be unethical? For example, companies are legally permitted to "tap" company telephone lines and monitor E-mail. Is it unethical for a safety and resource control professional to monitor a telephone call from the nurses' station to a physician's office?

The Americans with Disability Act (ADA) requires that the medical files and personnel files be separated and medical files be maintained in a confidential manner. Can information contained in these files be used for alternative purposes? For example, the safety and resource control professional identifies a period of alcohol rehabilitation in the medical file and specifically places the employee in an area where it is highly likely the employee will resume drinking and lose all protections under the ADA.

Companies often perform a wide variety of tests or examinations on employees after making an offer of employment to determine their qualifications for the job function. Can tests provide information beyond the specific issues in which the test was designed? Can the test itself be unethical or illegal? Would you test an employee's DNA? Can this information be used for other purposes? For example, some companies use a demanding lifting test that requires extensive upper body strength. These tests have been found to be discriminatory in design.

Job or performance appraisals are, in essence, a subjective or quasi-objective method of documenting employees performance and are often used for feedback and documentation purposes. Can job performance appraisals be manipulated for other purposes? For example, can the boss purposefully provide a low subjective performance appraisal to lay the groundwork for an upcoming demotion?

Are lockers, and the contents within, accessible to search by the company? By law enforcement? Under most circumstances, if the locker and lock are provided by the company, a search is permissible. However, can locker searches, even if legal, create unethical situations? For example, what about a boss who searches a specific employee locker in search of photographs of his or her spouse.

Under the Occupational Safety and Health Act (OSH) and other laws and regulations, employees have the right to view personal records under several standards and laws. Can review of records create ethical issues through the method of review? Through the extent of review? For example, a employee requests the right to review his medical file and is provided access to all employee medical files. Being nosey, this employee reviews another employee's file and finds out that the employee is human immunodeficiency virus (HIV) positive. The employee tells other employees in the facility.

Can a company search an employee's lunch box? Conduct a "pat-down" search? Order a "strip search"? Depending on the company, different types of searches may or may not be permitted. However, does conducting any type of search invade the individual's privacy rights? Can legal and ethical issues evolve because of searches?

Under several laws and regulatory schemes, employees are protected for reporting illegal activities to the governing agency. For example, employees are protected for reporting illegal dumping activities by a company. Safety and resource control professionals should exercise extreme caution in situations involving whistleblowing due to the civil, and often criminal, repercussions. (A good example is the recent tobacco litigation.) Can ethical issues as well as legal issues be involved in whistleblowing situations? For example, OSHA conducts an inspection based on a complaint. Although the company does not know the name of the employee filing the complaint, the boss wants all employees in a specific department terminated. What are you going to do?

Although not common in the United States, the use of monies to acquire certain concessions in other countries is common practice. In the United States, bribes, kickbacks, and payoffs are usually illegal. Most corporate policies forbid such practices. However, ethical, as well as legal, issues can result. For example, your company is performing a large project in a foreign country requiring the hiring of numerous workers. The safety and resource control professional is required to hire through a specific governmental agency, however, the director of the organizations requires a kickback on each worker placed. Without the workers, your company will lose the project. What are you going to do?

Is it deceptive to run an advertisement for a job in full knowledge that the job is already slated to be filled by the boss's brother-in-law? Is this deceptive or job a legal tactic? Does the company say it does not discriminate and then covertly circumvents the written policy in hiring, promotion, or other areas? Although there are many illegalities in this area, this area is also a minefield for ethical issues. For example, the safety and resource control professional possesses all policies and procedures to perform an unbias and nondiscriminatory

SAMPLE: Corporate Compliance Program Checklist

To assess whether your company is in need of a comprehensive compliance program in the area of safety and resource control and other areas of potential risk, the following list of questions is provided in order to assess your current position. Every "no" answer should send a signal that a potential risk is at hand and a program is needed.

1. Does the board of directors place a high priority on human resources, safety and health, environmental, resource control, discrimination areas, and other regulatory compliance requirements?
2. Has the company adopted policies with regard to compliance with governmental regulations and other laws having a direct bearing on the operations?
3. Has the company established and published a code of conduct and distributed copies to employees?
4. Has your company employed an individual(s) who will be directly responsible for human resources, safety and health and government compliance? Are these individuals properly educated and prepared to manage these functions?
5. Has the company formally developed programs to involve employees such as safety committees?
6. Does your company possess all the necessary resources to effectively develop and maintain an affective compliance programs?
7. Are your required personnel, safety and health, environmental and other compliance programs in writing?
8. Are corporate officers and managers and supervisors sensitive to the importance of government compliance?
9. Are your employees involved in your safety and resource control and related efforts?
10. Are your corporate officers and managers committed to your safety and resource control compliance efforts? Do they provide the necessary resources and staffing, to effectively perform the function successfully?

11. Does the company conduct periodic compliance and other legal audits to detect compliance failures? Are deficiencies or failures corrected in a timely manner?
12. Has the company established a "hotline" or other mechanisms through which to facilitate reporting of human resource, safety and resource control, and environmental compliance failures?
13. Are all employees properly trained in the required aspects of your compliance programs?
14. Is your training properly documented? That is, will your documentation prove beyond a shadow of a doubt that a particular employee was trained in a particular regulatory requirement?
15. Do you have a new employee orientation program in place?
16. Does the orientation program for new employees include review of personnel policies, safety and resource control policies, codes of conduct, other policies, and procedures?
17. Are employees provided additional training? Other on-the-job training?
18. Does your company conduct compliance training sessions to sensitive managers and rank-and-file employees to their legal responsibilities?
19. Does your company provide information and assistance to employees regarding their rights and responsibilities under the individual state's workers' compensation laws?
20. Does your company communicate human resource, safety and resource control compliance issues to employees with posting, newsletters, brochures, and manuals?
21. Does your company go beyond the bare bones compliance requirements to create an appropriate work environment?
22. Does your company keep up with the new and revised laws, standards, and other regulatory compliance requirements?
23. Does your company appropriately discipline employees for failure to follow rules, policies, and regulations? Is this discipline fair and consistent?
24. Is creating a good working environment a high priority for your officers and directors?
25. Is your company proactive in the areas of human resources, safety and resource control, and environment?

selection process and is informed prior to hiring that a specific male candidate must be selected. What would you do?

Although every safety and resource control professional strives to achieve a program that is beneficial to the company as well as all employees, sometimes specific policies, programs, or procedures can result in an detrimental situation for an individual employee. For example, the safety and resource control professional negotiates a better health insurance program for the company, which provides better benefits at a lower cost with a quality company. However, due to a specific pre-existing condition, coverage is not provided for a one employee's child.

Is it "stealing" to take pens, notepads, and other company property home? Is it unethical to "borrow" company supplies for use at home? What is your company's policy in this area? How is the policy enforced uniformly and fairly?

In developing a corporate code of conduct for use within a company, special care should be provided in order to structure the code properly to encompass a broad spectrum of potential risks, while also providing guidance to employees as to the expected behaviors. Some of the items to be considered include the following:

1. *Content of a corporate code of conduct.* The most common laws covered in the codes of conduct include labor law, antitrust law, business ethics, conflicts of interest, corporate political activity, environmental law, safety and health laws, employee relations law, securities laws, etc. Special care should be provided in the area of individual and personal rights.
2. *Distribution to officers, directors, and employees.* Most companies distribute their codes of conduct to directors, officers, and employees when they join the company and on a yearly basis. the company may want to document that the director, officer, and employee have read, understood, and will adhere to the Code.
3. *Provide training and education to promote compliance.* In-house seminars can be targeted to small groups of key managers to sensitize them to their legal responsibilities. Some companies use guidebooks as memos to communicate the importance of compliance to their employees. Newsletters and memos can be distributed to managers on a periodic basis to remind them about their legal responsibilities, advise them of the developments of the law, and give them preventive law tips.

4. *Enforcing the code of conduct.* Compliance programs should include disciplinary procedures to punish violations of Code of Conduct. Compliance programs are not effective if they are not enforced. Sanctions for violation can include verbal warnings, written warnings, suspension, demotion, discharge, and referral to law enforcement agencies. In addition to sanctions, the disciplinary procedures should include provisions for protecting whistle blowers and for investigation allegations of illegal conduct.
5. *Monitor compliance through legal audits.* Legal audits can include, but are not limited to, the following activities:
   a. Assemble legal audit team.
   b. Educate and train legal audit team.
   c. Assign legal team responsibilities.
   d. Develop a site-specific audit instrument.
   e. Conduct site visitation and facility inspection.
   f. Conduct employee interviews.
   g. Conduct records search.
   h. Develop and use of employee questionnaires.
   i. Review record retention procedures and policies.
   j. Develop an audit report.
   k. Present audit report to board of directors and officers.
   l. The legal audit mechanism, like the internal safety and health audit discussed above, produces documentation that may be the subject of discovery requests in civil or criminal litigation. This type of evaluation may also produce sensitive data that the company seeks to keep confidential. Extreme caution should be exercised to preserve privilege and confidentiality.

In summary, safety and resource control professionals should assess their potential risks and balance these risks versus the possible benefits and possible detriments. With the appropriate preparation, the vast majority of potential risks can be properly managed to safeguard not only your company, but also your personnel and yourself. Identifying the risk, assessing the risk, and adopting a proactive approach before something happens can minimize or eliminate the impact of the risk or even the risk itself.

CHAPTER 10

THE COMPLIANCE CIRCLE

> While all other sciences have advanced, that of government is at a standstill—little better understood, little better practiced now than three or four thousand years ago.
>
> —John Adams

> It gets harder and harder to support the government in a manner to which it has become accustomed.
>
> —Anonymous

Safety and resource control professionals often have direct responsibility or, at a minimum, oversight authority when your organization interacts with various governmental agencies. It is vitally important that safety and resource control professionals, thoroughly understanding the scope and magnitude of this important responsibility. Normally, the primary government agencies that oversee most public and private sector safety and resource control professionals include the federal or state agencies responsible for workplace safety and health, the federal or state agencies responsible for environmental protection, and the federal or state agencies responsible for the enforcement of labor and employment laws. Safety and resource control professionals should be aware that virtually all agencies have specific rules and regulations, and it is important that the safety and resource control professional become familiar with these rules before an inspection or other compliance activity.

The primary government agency that safety and resource control professionals will interact with on virtually a daily basis is the Occupational Safety and Health Administration (OSHA). Safety and resource control professionals must possess a working knowledge not only of the numerous standards but of the "workings" of the agency, the history of the agency, and the direction in which the agency may travel in the future in order to complete the "compliance circle." The Occupational Safety and Health (OSH) Act of 1970 is recognized as the birthplace of the modern safety and resource control profession.

Before the Federal Occupational Safety and Health Act (hereinafter referred to as the OSH Act) of 1970 was enacted, safety and health compliance was limited to specific industry safety and health laws and laws that governed federal contractors. It was during this period before the enactment of the OSH Act that Congress gradually began to regulate safety and health in the American workplace through such laws as the Walsh–Healey Public Contracts Act of 1936, the Labor Management Relations Act (Taft–Hartley Act) of 1947, the Coal Mine Safety Act of 1952, and the McNamara–O'Hara Public Service Contract Act of 1965.

With the passage of the controversial OSH Act in 1970, federal and state governmental agencies now became actively involved in managing health and safety in the private sector workplace. Employers were placed on notice that unsafe and unhealthful conditions/acts would no longer be permitted to endanger the health, and often the lives, of American workers. In many circles, the Occupational Safety and Health Administration (OSHA) became synonymous with the "safety police," and employers were often forced, under penalty of law, to address safety and health issues in their workplace.

Today, the OSH Act itself is virtually unchanged since its 1970 roots, and the basic methods for enforcement, standards development and promulgation, as well as adjudication are virtually unchanged. OSHA has, however, added many new standards during the past 30 years based primarily on the research conducted by the National Institute for Occupational Safety and Health (NIOSH) and recommendations from labor and industry. Moreover, the Occupational Safety and Health Review Commission (OSHRC) has been very active in resolving many disputed issues and clarifying the law as it stands.

There is trend within Congress, industry, and labor that, in order to achieve the ultimate goal of reducing workplace injuries, illnesses, and fatalities, additional changes are needed to the OSH Act and the structure of OSHA. Change is likely to be forthcoming, but it will be based upon the learning from past mistakes and on the resolution of issues in order to achieve our ultimate goal of safe and healthful workplace for all.

The OSH Act covers virtually every American workplace that employs one or more employees and engages in a business that in any way affects interstate commerce.[1] The OSH Act covers employment in every state, the District of Columbia, Puerto Rico, Guam, the Virgin Islands, American Samoa, and the Trust Territory of the Pacific Islands.[2] The OSH Act does not cover employees in situations in which other state or federal agencies have jurisdiction that requires the agencies to prescribe or enforce their own safety and health regulations.[3] The OSH Act also exempts residential owners who employ people for ordinary domestic tasks, such as cooking, cleaning, and child care.[4] It also does not cover federal,[5] state, and local governments[6] or Native American reservations.[7]

The OSH Act does require every employer engaged in interstate commerce furnish employees "a place of employment ... free from recognized hazards that are causing, or are likely to cause, death or serious harm."[8] To help employers create and maintain safe working environments, and to enforce laws and regulations that ensure safe and healthful work environments, Congress provided for the creation of the OSHA, to act as a new agency under the direction of the Department of Labor.

Today, OSHA is one of the most widely known and powerful enforcement agencies. It has been granted broad regulatory powers to promulgate regulations and standards, investigate and inspect, issue citations, and propose penalties for safety violations in the workplace.

The OSH Act also established an independent agency to review OSHA citations and decisions, the OSHRC. The OSHRC is a quasi-judicial and independent administrative agency composed of three commissioners appointed by the president who serve staggered 6-year terms. The OSHRC has the power to issue orders, uphold, vacate, or modify OSHA citations and penalties and to direct other appropriate relief and penalties.

The educational arm of the OSH Act is the NIOSH, which was created as a specialized educational agency of the existing National Institutes of Health (NIH). NIOSH conducts occupational safety and health research and develops criteria for new OSHA standards. NIOSH can conduct workplace inspections, issue subpoenas, and question employees and employers, but it does not have the power to issue citations or penalties.

As permitted under the OSH Act, OSHA encourages individual states to take responsibility for OSHA administration and enforcement within their own respective boundaries. Each state possesses the ability to request and be granted the right to adopt state safety and health regulations and enforcement mechanisms.[9] For a state plan to be placed into effect, the state must first develop and submit its proposed program to the Secretary of Labor for review and approval. The Secretary must certify that the state plan's standards are "at least as effective" as the federal standards and that the state will devote adequate resources to administering and enforcing standards.[10]

In most state plans, the state agency has developed more stringent safety and health standards than OSHA[11] and, in most cases, has developed more stringent enforcement schemes.[12] The Secretary of Labor has no statutory authority to reject a state plan if the proposed standards or enforcement scheme are more strict than the OSHA standards but can reject the state plan if the standards are below the minimum limits set under OSHA standards.[13] These states are known as "state plan" states and territories.[14] (As of this writing, there were 21 states and two territories with approved and functional state plan programs.[15]) Employers in state plan states and territories must comply with their state's regulations; federal OSHA plays virtually no role in direct enforcement.

OSHA does, however, possess an approval and oversight role with regard to state plan programs. OSHA must approve all state plan proposals before enactment, and they maintain oversight authority to "pull the ticket" of any or all state plan programs at any time they are not achieving the identified prerequisites. Enforcement of this oversight authority was recently observed after the fire, resulting in several workplace fatalities at the Imperial Foods facility in Hamlet, North Carolina. After this incident, federal OSHA assumed jurisdiction and control over the state plan program in North Carolina and made significant modifications to this program before returning the program to state control.

The OSH Act requires that a covered employer comply with specific occupational safety and health standards and all rules, regulations, and orders issued pursuant to the OSH Act that apply to the workplace.[16] The OSH Act also requires that all standards be based on research, demonstration, experimentation, or other appropriate information.[17] The Secretary of Labor is authorized under the Act to "promulgate, modify, or revoke any occupational safety and health standard,"[18] and the OSH Act describes the procedures that the Secretary must follow when establishing new occupational safety and health standards.[19]

The OSH Act authorizes three ways to promulgate new standards. From 1970 to 1973, the Secretary of Labor was authorized in Section 6(a) of the Act[20] to adopt national consensus standards and establish federal safety and health standards without following lengthy rulemaking procedures. Many of the early OSHA standards were adapted mainly from other areas of regulation, such as the National Electric Code and American National Standards Institute (ANSI) guidelines. However, this promulgation method is no longer in effect.

The usual method of issuing, modifying, or revoking a new or existing OSHA standard is set out in Section 6(b) of the OSH Act and is known as informal rulemaking. It requires notice to interested parties, through subscription in the *Federal Register* of the proposed regulation and standard, and provides an opportunity for comment in a nonadversarial administrative hearing.[21] The proposed standard can also be advertised through magazine articles and other publications, thus informing interested parties of the proposed standard and regulation. This method differs from the requirements of most other administrative agencies that follow the Administrative Procedure Act[22] in that the OSH Act provides interested persons an opportunity to request a public hearing with oral testimony. It also requires the Secretary of Labor to publish in the *Federal Register* a notice of the time and place of such hearings.

Although not required under the OSH Act, the Secretary of Labor has directed, by regulation, that OSHA follow a more rigorous procedure for comment and hearing than other administrative agencies.[23] Upon notice and request for a hearing, OSHA must provide a hearing examiner in order to listen to any oral testimony offered. All oral testimony is preserved in a verbatim transcript. Interested persons are provided an opportunity to cross-examine OSHA representatives or others on critical issues. The Secretary must state the reasons for the action to be taken on the proposed standard, and the statement must be supported by substantial evidence in the record as a whole.

The Secretary of Labor has the authority not to permit oral hearings and to call for written comment only. Within 60 days after the period for written comment or oral hearings has expired, the Secretary must decide whether to adopt, modify, or revoke the standard in question. The Secretary can also decide not to adopt a new standard. The Secretary must then publish a statement of the reasons for any decision in the *Federal Register*. OSHA regulations further mandate that the Secretary provide a supplemental statement of significant issues in the decision. Safety and health professionals should be aware that the standard as adopted and published in the *Federal Register* may be different from the proposed standard. The Secretary is not required to reopen hearings when the adopted standard is a logical outgrowth of the proposed standard.[24]

The final method for promulgating new standards, and the one most infrequently used, is the emergency temporary standard permitted under Section 6(c).[25] The Secretary of Labor may establish a standard immediately if it is determined that employees are subject to grave danger from exposure to substances or agents known to be toxic or physically harmful, and that an emergency standard would protect the employees from the danger. An emergency temporary standard becomes effective on publication in the *Federal Register* and may remain in effect for 6 months. During this 6-month period, the Secretary must adopt a new permanent standard or abandon the emergency standard.

Only the Secretary of Labor can establish new OSHA standards. Recommendations or requests for an OSHA standard can come from any interested person or organization, including employees, employers, labor unions, environmental groups, and others.[26] When the Secretary receives a petition to adopt a new standard or to modify or revoke an existing standard, he or she usually forwards the request to NIOSH and the National Advisory Committee on Occupational

Safety and Health (NACOSH)[27] or the Secretary may use a private organization such as ANSI for advice and review.

The OSH Act requires that an employer maintain a place of employment free from recognized hazards that are causing or are likely to cause death or serious physical harm, even if there is no specific OSHA standard addressing the circumstances. Under Section 5(a)(1), known as the "general duty clause," an employer may be cited for a violation of the OSH Act if the condition causes harm or is likely to cause harm to employees, even if OSHA has not promulgated a standard specifically addressing the particular hazard. The general duty clause is a catch-all standard encompassing all potential hazards that have not been specifically addressed in the OSHA standards. For example, if a company is cited for an ergonomic hazard and there is no ergonomic standard to apply, the hazard will be cited under the general duty clause.

Safety and resource control professionals should take a proactive approach in maintaining their competency in this expanding area of OSHA regulations. As noted previously, the first notice of any new OSHA standard, modification of an existing standard, revocation of a standard, or emergency standard must be published in the *Federal Register*. Safety and resource control professionals can use the *Federal Register*, or professional publications that monitor OSHA standards, to track the progress of proposed standards. With this information, safety and resource control professionals can provide testimony to OSHA when necessary, prepare their organizations for acquiring resources and personnel necessary to achieve compliance, and get a head start on developing compliance programs to meet requirements in a timely manner.

The OSH Act provides for a wide range of penalties, from a simple notice with no fine to criminal prosecution. The Omnibus Budget Reconciliation Act of 1990 multiplied maximum penalties sevenfold. Violations are categorized, and penalties may be assessed as outlined in Table 10.1.

Each alleged violation is categorized and the appropriate fine issued by the OSHA area director. It should be noted that each citation is separate and may carry with it a monetary fine. The gravity of the violation is the primary factor in determining penalties.[28] In assessing the gravity of a violation, the compliance officer or area

TABLE 10.1 Violation and Penalty Schedule

Penalty	Old Penalty Schedule (in dollars)	New Penalty Schedule (1990) (in dollars)
De minimis notice	0	0
Nonserious	0–1000	0–7000
Serious	0–1000	0–7000
Repeat	0–10,000	0–70,000
Willful	0–10,000	25,000 minimum 70,000 maximum
Failure to abate notice	0–1000 per day	0–7000 per day
New posting penalty		0–7000

director must consider (1) the severity of the injury or illness that could result, and (2) the probability that an injury or illness *could* occur as a result of the violation.[29] Specific penalty assessment tables assist the area director or compliance officer in determining the appropriate fine for the violation.[30]

After selecting the appropriate penalty table, the area director or other official determines the degree of probability that the injury or illness will occur by considering the following[31]:

1. The number of employees exposed.
2. The frequency and duration of the exposure.
3. The proximity of employees to the point of danger.
4. Factors such as the speed of the operation that require work under stress.
5. Other factors that might significantly affect the degree of probability of an accident.

OSHA has defined a serious violation as "an infraction in which there is a substantial probability that death or serious harm could result ... unless the employer did not or could not with the exercise of reasonable diligence, know of the presence of the violation."[32] Section 17(b) of the OSH Act requires that a penalty of up to $7,000 be assessed for every serious violation cited by the compliance officer.[33] In assembly-line enterprises and manufacturing facilities with duplicate operations, if one process is cited as possessing a serious violation, it is possible that each of the duplicate processes or machines may be cited for the same violation. Thus, if a serious violation is found in one machine and there are many other identical machines in the enterprise, a very large monetary fine for a single serious violation is possible.[34]

Currently the greatest monetary liabilities are for repeat violations, willful violations, and failure to abate cited violations.

A repeat violation is a second citation for a violation that was cited previously by a compliance officer. OSHA maintains records of all violations and must check for repeat violations after each inspection.

A willful violation is the employer's purposeful or negligent failure to correct a known deficiency. This type of violation, in addition to carrying a large monetary fine, exposes the employer to a charge of an egregious violation and the potential for criminal sanctions under the OSH Act or state criminal statutes if an employee is injured or killed as a direct result of the willful violation.

Failure to abate a cited violation has the greatest cumulative monetary liability of all. OSHA may assess a penalty of up to $1000 per day per violation for each day in which a cited violation is not brought into compliance.

In assessing monetary penalties, the area or regional director must consider the good faith of the employer, the gravity of the violation, the employer's past history of compliance, and the size of the employer. In addition to the potential civil or monetary penalties that could be assessed, OSHA regulations may be used as evidence in negligence, product liability, workers' compensation, and other actions involving employee safety and health issues.[36] OSHA standards and regulations are the baseline requirements for safety and health that must be met, not only to achieve compliance with the OSHA regulations, but also to safeguard an organization against other potential civil actions.

The OSH Act provides for criminal penalties under four circumstances.[37] First, anyone inside or outside the Department of Labor or OSHA who gives advance notice of an inspection, without authority from the Secretary, may be fined up to $1000 or imprisoned for up to 6 months, or both. Second, any employer or person who intentionally falsifies statements or OSHA records that must be prepared, maintained, or submitted under the OSH Act may if found guilty be fined up to $10,000 or imprisoned for up to 6 months, or both. Third, any person responsible for a violation of an OSHA standard, rule, order, or regulation, who causes the death of an employee may, upon conviction, be fined up to $10,000 or imprisoned for up to 6 months, or both. If convicted for a second violation, punishment may be a fine of up to $20,000 or imprisonment for up to 1 year, or both.[38] Finally, if an individual is convicted of forcibly resisting or assaulting a compliance officer or other Department of Labor personnel, a fine of $5000 or 3 years in prison, or both, can be imposed. Any person convicted of killing a compliance officer or other OSHA or Department of Labor personnel acting in his or her official capacity may be sentenced to prison for any term of years or life.

OSHA does not have authority to impose criminal penalties directly, instead, it refers cases for possible criminal prosecution to the U.S. Department of Justice. Criminal penalties must be based on violation of a specific OSHA standard; they may not be based on a violation of the general duty clause. Criminal prosecutions are conducted like any other criminal trial, with the same rules of evidence, burden of proof, and rights of the accused. A corporation may be criminally liable for the acts of its agents or employees.[39] The statute of limitations for possible criminal violations of the OSH Act, as for other federal noncapital crimes, is 5 years.[40]

Under federal criminal law, criminal charges may range from murder to manslaughter to conspiracy. Several charges may be brought against an employer for various separate violations under one federal indictment.

The OSH Act provides for criminal penalties of up to $10,000 or imprisonment for up to 6 months, or both. A repeated willful violation causing an employee death can double the criminal sanction to a maximum of $20,000 or 1 year of imprisonment, or both. Given the increased use of criminal sanctions by OSHA in recent years, personnel managers should advise their employers of the potential for these sanctions being used when the safety and health of employees are disregarded or placed on the back burner.

Criminal liability for a willful OSHA violation can attach to an individual or a corporation. In addition corporations may be held criminally liable for the actions of their agents or officials.[41] Corporate officials may also be subject to criminal liability under a theory of aiding and abetting the criminal violation in their official capacity with the corporation.[42]

Safety and resource control professionals should exercise extreme caution when faced with an on-the-job fatality. The potential for criminal sanctions and criminal prosecution is substantial if a willful violation of a specific OSHA standard is directly involved in the death. The OSHA investigation may be conducted from a criminal perspective in order to gather and secure the appropriate evidence to later pursue criminal sanctions.[43] Safety and resource control professionals facing a workplace fatality investigation should address the OSHA investigation with legal counsel present and reserve all rights guaranteed under the U.S. Constitution.[44] Obviously, under no circumstances should a safety and resource control professional condone or attempt to conceal facts or evidence that consist of a cover-up.

OSHA performs all enforcement functions under the OSH Act. Under Section 8(a) of the Act, OSHA compliance officers have the right to enter any workplace of a covered employer without delay, inspect and investigate a workplace during regular hours and at other reasonable times and obtain an inspection warrant if access to a facility or operation is denied.[45] Upon arrival at an inspection site, the compliance officer is required to present his or her credentials to the owner or designated representative of the employer before starting the inspection. The employer representative and an employee or union representative, or both, may accompany the compliance officer on the inspection. Compliance officers can question the employer and employees and inspect required records, such as the OSHA Form 200, which records injuries and illnesses.[46] Most compliance officers cannot issue on-the-spot citations; they only have authority to document potential hazards and report or confer with the OSHA area director before issuing a citation.

A compliance officer nor any other employee of OSHA may not provide advance notice of the inspection under penalty of law.[47] The OSHA area director is, however, permitted to provide notice under the following circumstances[48]:

1. In cases of apparent imminent danger, to enable the employer to correct the danger as quickly as possible.
2. When the inspection can most effectively be conducted after regular business hours or where special preparations are necessary.
3. To ensure the presence of employee and employer representatives or appropriate personnel needed to aid in inspections.
4. When the area director determines that advance notice would enhance the probability of an effective and through inspection.

Compliance officers can also take environmental samples and obtain photographs related to the inspection. Compliance officers can use other "reasonable investigative techniques," including personal sampling equipment, dosimeters, air sampling badges, and other equipment.[49] Compliance officers must, however, take reasonable precautions when using photographic or sampling equipment to avoid creating hazardous conditions (i.e., a spark-producing camera flash in a flammable area) or disclosing a trade secret.[50]

An OSHA inspection has four basic components: (1) the opening conference, (2) the walk-through inspection, (3) the closing conference, and (4) the issuance of citations, if necessary. In the opening conference, the compliance officer may explain the purpose and type of inspection to be conducted, request records to be evaluated, question the employer, ask for appropriate representatives to accompany him or her during the walk-through inspection, an ask additional questions or request more information. The compliance officer may, but is not required to, provide the employer with copies of the applicable laws and regulations governing procedures and health and safety standards. The opening conference is usually brief and informal, its primary purpose is to establish the scope and purpose of the walk-through inspection.

After the opening conference and review of appropriate records, the compliance officer, usually accompanied by a representative of the employer and a representative of the employees, conducts a physical inspection of the facility or worksite.[51] The general purpose of this walk-through inspection is to determine whether the facility or worksite complies with OSHA standards. The compliance officer must identify potential safety and health hazards in the workplace, if any, and document them to support issuance of citations.[52]

When the walk-through inspection is completed, the compliance officer usually conducts an informal meeting with the employer or the employer's representative to "informally advise (the employer) of any apparent safety or health violations disclosed by the inspection."[53] The compliance officer informs the employer of the potential hazards observed and indicates the applicable section of the standards allegedly violated, advises that citations may be issued, and informs the employer or representative of the appeal process and rights.[54] The compliance officer also advises the employer that the OSH Act prohibits discrimination against employees or others for exercising their rights.[55]

When the compliance officer completes the inspection, the information is provided to the Secretary of Labor, to decide whether a citation should be issued, compute any penalties to be assessed; and set the abatement date for each alleged violation. The area director, under authority from the Secretary, must issue the citation with "reasonable promptness."[56] Citations must be issued in writing and must describe with particularity the violation alleged, including the relevant standard and regulation. There is a 6-month statute of limitations and the citation must be issued or vacated within this time period. OSHA must serve notice of any citation and proposed penalty by certified mail, unless there is personal service, to an agent or officer of the employer.[57]

After the citation and notice of proposed penalty is issued, but before the notice of contest by the employer is filed, the employer may request an informal conference with the OSHA area director. The general purpose of the informal conference is to clarify the basis for the citation, modify abatement dates or proposed penalties, seek withdrawal of a cited item, or otherwise attempt to settle the case. This conference, as its name implies, is an informal meeting between the employer and OSHA. Employee representatives must have an opportunity to participate if they so request. Safety and resource control professionals should note that the request for an informal conference does not "stay" (delay) the 15-working-day period to file a notice of contest to challenge the citation.[58]

Under the OSH Act, an employer, employee, or authorized employee representative (including, a labor organization) is given 15 working days from when the citation is issued to file a "notice of contest." If a notice of contest is not filed within 15 working days, the citation and proposed penalty become a final order of the OSHRC and are not subject to review by any court or agency. If a timely notice of contest is filed in good faith, the abatement requirement is tolled (temporarily suspended or delayed) and a hearing is scheduled. The employer also has the right to file a petition for modification of the abatement period (PMA) if the employer is unable to comply with the abatement period provided in the citation. If OSHA contests the PMA, a hearing is scheduled to determine whether the abatement requirements should be modified.

When the notice of contest by the employer is filed, the Secretary must immediately forward the notice to the OSHRC, which then schedules a hearing before its administrative law judge (ALJ). The Secretary of Labor is labeled the "complainant," and the employer the "respondent." The ALJ may affirm, modify, or vacate the citation, any penalties, or the abatement date. Either party can appeal the ALJ's decision by filing a petition for discretionary review (PDR). In addition, any member of the OSHRC may "direct review" of any decision by an ALJ, in whole or in part, without a PDR. If a PDR is not filed and no member of the OSHRC directs a review, the decision of the ALJ becomes final in 30 days. Any party may appeal a final order of the OSHRC by filing a petition for review in the U.S. Court of Appeals for the circuit in which the violation is alleged to have occurred or in the U.S. Court of Appeals for the District of Columbia Circuit. This petition for review must be filed within 60 days from the date of the OSHRC's final order.

The following checklist is intended to assist safety and resource control professionals prepare for an OSHA inspection:

1. Assemble a team from the management group and identify specific responsibilities, in writing, for each team member. The team members should be given appropriate training and education and should include, but not be limited to the following[59]:
   a. An OSHA inspection team coordinator.
   b. A document control individual.
   c. Individuals to accompany the OSHA inspector.
   d. An accident investigation team leader (where applicable).

  e. A notification person.
  f. A legal advisor (where applicable).
  g. A law enforcement coordinator (where applicable).
  h. A photographer.
  i. An industrial hygienist.
  j. Media coordinator.
2. Decide on and develop a company policy and procedures to provide guidance to the OSHA inspection team.
3. Prepare an OSHA inspection kit, including all equipment necessary to properly document all phases of the inspection. The kit should include equipment such as a camera (with extra film and batteries), a tape player (with extra batteries), a video camera, pads, pens, and other appropriate testing and sampling equipment (e.g., a noise level meter, an air sampling kit).
4. Prepare basic forms to be used by the inspection team members during and following the inspection.
5. When notified that an OSHA inspector has arrived, assemble the team members along with the inspection kit.
6. Identify the inspector. Check the inspector's credentials and determine the reason for, and type of, inspection to be conducted.
7. Confirm the reason for the inspection with the inspector (targeted, routine inspection, accident, or in response to a complaint?)
  a. For a random or target inspection:
     - Did the inspector check the OSHA 200 Form?
     - Was a warrant required?
  b. For an employee complaint inspection:
     - Did inspector have a copy of the complaint? If so, obtain a copy.
     - Do allegations in the complaint describe an OSHA violation?
     - Was a warrant required?
     - Was the inspection protested in writing?

c. For an accident investigation inspection:
   - How was OSHA notified of the accident?
   - Was a warrant required?
   - Was the inspection limited to the accident location?

d. If a warrant is presented:
   - Were the terms of the warrant reviewed by local counsel?
   - Did the inspector follow the terms of the warrant?
   - Was a copy of the warrant acquired?
   - Was the inspection protested in writing?

8. The opening conference
   a. Who was present?
   b. What was said?
   c. Was the conference taped or otherwise documented?
9. Records
   a. What records were requested by the inspector?
   b. Did the document control coordinator number the photocopies of the documents provided to the inspector?
   c. Did the document control coordinator maintain a list of all photocopies provided to the inspector?
10. Facility inspection
    a. What areas of the facility were inspected?
    b. What equipment was inspected?
    c. Which employees were interviewed?
    d. Who was the employee or union representative present during the inspection?
    e. Were all the remarks made by the inspector documented?
    f. Did the inspector take photographs?
    g. Did a team member take similar photographs?

There is no replacement for a well-managed safety and resource control program. Safety and resource control professionals should ensure that their upper level management team realizes that they cannot get by on a shoestring safety and resource control program. Every aspect of the program is important, including preparing for an

OSHA inspection. This checklist was provided as an example of what can be done prior to an OSHA inspection.

When a compliance officer or other Department of Labor representative enters a facility to perform an inspection, the employer has certain rights. It is entitled to know the purpose of the inspection, such as whether it is based on an employee complaint or is a routine inspection. The employer also has the right to accompany the compliance officer during the inspection. This can be helpful or harmful, helpful in the sense that the employer can avoid certain areas, but harmful if a major violation is found and the employer in trying to explain, talks itself into more trouble and ends up with a higher fine and a more serious violation. (See Appendix C for complete listing of Rights during an OSHA inspection.)

Under Section 9(a) of the OSH Act, if the Secretary of Labor believes that an employer "has violated a requirement of Section 5 of the Act, of any standard, rule or order promulgated pursuant to Section 6 of this Act, or of any regulations prescribed pursuant to this Act, he shall within reasonable promptness issue a citation to the employer."[60] "Reasonable promptness" has been defined to mean within 6 months from the occurrence of a violation.[61]

Section 9(a) also requires that citations be in writing and "describe" with particularity the nature of the violation, including a reference to the provision of the Act, standard, rule, regulation, or order alleged to have been violated."[62] The OSHRC has adopted a fair notice test that is satisfied if the employer is notified of the nature of the violation, the standard allegedly violated, and the location of the alleged violation.[63]

The OSH Act does not specifically provide for a method of service for citations. Section 10(a) authorizes service of notice of proposed penalties by certified mail, and in most instances the written citations are attached to the penalty notice.[64] Regarding the proper party to be served, the OSHRC has held that service is proper if it "is reasonably calculated to provide an employer with knowledge of the citation and notification of proposed penalty and an opportunity to determine whether to contest or abate."[65]

Under Section 10 of the Act, once a citation is issued, the employer, any employee, or any authorized union representative has 15 working days to file a notice of contest.[66] If the employer does not contest the violation, abatement date, or proposed penalty, the citation becomes final and not subject to review by any court or agency. If a timely notice of contest is filed in good faith, the abatement requirement is tolled and a hearing is scheduled. An employer may contest any part or all of the citation, proposed penalty, or abatement date. Employee contests are limited to the reasonableness of the proposed abatement date. Employees also have the right to elect party status after an employer has filed a notice of contest.

Safety and resource control pofessionals may also file a PMA if it cannot comply with any abatement that has become a final order. If the Secretary of Labor or an employer contests the PMA, a hearing is held to determine whether any abatement requirement, even if part of an uncontested citation, should be modified.[67]

The notice of contest does not have to be in any particular form and is sent to the area director who issued the citation. The area director must forward the notice to the OSHRC, and the OSHRC must docket the case for hearing.

After pleading, discovery, and other preliminary matters, a hearing is scheduled before an ALJ. Witnesses testify and are cross-examined under oath, and a verbatim transcript is made. The Federal Rules of Evidence apply.[68]

Following closure of the hearing, parties may submit briefs to the ALJ. The ALJ's decision contains findings of fact and conclusions of law and affirms, vacates, or modifies the citation, proposed penalty, and abatement requirements. The ALJ's decision is filed with the OSHRC and may be directed for review by any OSHRC member *sua sponte* or in response to a party's petition for discretionary review. Failure to file a petition for discretionary review precludes subsequent judicial review.

The Secretary of Labor has the burden of proving the violation. The hearing is presided over by an ALJ and he or she renders a decision either affirming, modifying, or vacating the citation, penalty, or abatement date. The ALJ's decision then automatically goes before the OSHRC. The aggrieved party may file a petition requesting that the ALJ's decision be reviewed, but even without this discretionary review, any OSHRC member may direct review of any part or all of the ALJ's decision. If no member of the OSHRC directs a review within 30 days, however, the ALJ's decision is final. Through either review route, the OSHRC may reconsider the evidence and issue a new decision.

OSHRC review is, but the factual determinations of the ALJ, especially regarding credibility findings, are often afforded great weight. Briefs may be submitted to the OSHRC, but oral argument although extremely rare is also within the OSHRC's discretion.

In this administrative phase of the Act's citation adjudication process, the employer's good faith, the gravity of the violation, the employer's past history of compliance, and the employer's size are all considered in the penalty assessment. The area director can compromise, reduce, or remove a violation. Many citations can be compromised or reduced at this stage.

Although the OSHRC's rules mandate the filing of a complaint by the Secretary and an answer by the employer, pleadings are liberally construed and easily amended. Approximately 90% of the cases filed are resolved without a hearing, either through settlement, withdrawal of the citation by the Secretary, or withdrawal of the notice of contest by the employer.[69]

As permitted under Section 11(a) of the Act, any person adversely affected by the OSHRC's final order may file, within 60 days from the decision, a petition for review in the U.S. Court of Appeals for the circuit in which the alleged violation had occurred or in the U.S. Court of Appeals for the District of Columbia Circuit.[70] Under Section 11(b), the Secretary may seek review only in the circuit in which the alleged violation occurred or where the employer has its principal office.[71]

The courts apply the substantial evidence rule to factual determinations made by the OSHRC and its ALJs, but courts vary on the degree of deference afforded the OSHRC's interpretations of the statutes and standards. The burden of proof is on the Secretary of Labor at this hearing.[72] The rules of civil procedure, rules of evidence, and all other legal requirements apply, as with any trial before the federal court.

Safety and resource control professionals should be aware of their rights and responsibilities under the law. Although they should not fear inspections by OSHA, they should prepare for them, to ensure that the legal rights of the employer are protected. Under most circumstances, simple communication with the OSHA inspector or area director can correct most difficulties during an inspection. Once a citation is issued, the safety and resource control professional should at least consider contesting the citation at the area director level in order to discuss reduction of monetary penalties. If an amicable solution cannot be reached at the regional level, safety and resource control professionals should be certain that the time limitations are met in order to preserve the right to appeal the decision. Meeting the specific time limitations set forth in the Act is of utmost importance. If the time limitation is permitted to lapse, the opportunity for appeal is lost.

Compliance with the numerous laws, standards, and requirements mandated by various governmental agencies has become an important part of most safety and resource control professional's job functions. Safety and resource control professionals should be aware of the numerous regulatory requirements from other federal agencies including, but not limited to, the Environmental Protection Agency (EPA), Food and Drug Administration (FDA), U.S. Department of Agriculture (USDA), Equal Employment Opportunity Commission (EEOC), as well as various state and local agencies which may be applicable to their workplace. Each of these agencies usually possesses written rules and administrative procedures which are required to be followed. Given the potential detrimental effects on your company or organization, safety and resource control professionals must become proficient in the regulations applicable to their organization and/or acquire appropriate assistance. Safety and resource control professionals usually possess direct or indirect responsibility for this important area where achieving and maintaining compliance can be an enormous task. Proper management of this important function by the safety and resource control professional will provide a multitude of benefits for the employees, management team and the company. Failure to manage this important function appropriately is a recipe for disaster.

ENDNOTES

1. Ibid, §1975.3(d).
2. Ibid, §652(7).

3. *See, for example*, Atomic Energy Act of 1954, 42 U.S.C. §2021.
4. 29 C.F.R. §1975(6).
5. 29 U.S.C.A. §652(5). (No coverage under OSH Act, when U.S. government acts as employer.)
6. Ibid.
7. *See, for example*, *Navajo Forest Prods. Indus.*, 8 O.S.H. Cases 2694 (OSH Rev. Comm'n 1980), affd, 692 F.2d 709, 10 O.S.H. Cases 2159.
8. 29 U.S.C.A. §654(a)(1).
9. In Section 18(b), the OSH Act provides that any state "which, at any time, desires to assume responsibility for development and the enforcement therein of occupational safety and health standards relating to any ... issue with respect to which a federal standard has been promulgated ... shall submit a state plan for the development of such standards and their enforcement."
10. 8 Ibid, §667(c). After an initial evaluation period of a least three years during which OSHA retains concurrent authority, a state with an approved plan gains exclusive suthority over standard setting, inspection procedures, and enforcement of health and safety issures covered under the state plan. *See Also Noonan v. Texaco*, 713 P.2d 160 (Wyo. 1986); Plans for the Development and Enforcement of State Standards, 29 C.F.R. §667(f) (1982) and §1902. 42(c) (1986). Although the state plan is implemented by the individual state, OSHA continues to monitor the program and may revoke the state authority if the state does not fulfill the conditions and assurances contained within the proposed plan.
11. Some states incorporate federal OSHA standards into their plans and add only a few of their own standards as a supplement. Other states, such as Michigan and California, have added a substantialnumber of separate and independently promulgated standards. *See generally* Employee Safety and Health Guide (CCH) §§5000–5840 (1987) (Compiling all state plans). Some states also add their own penalty structures. For example, under Arizona's plan, employers may be fined up to $150,000 and sentenced to $1\frac{1}{2}$ years in prison for knowing violations of state standards that cause death to an employee and may also have to pay $25,000 in compensation to the victim's family. If the employer is a corporation, the maximum fine is $1 million. *See* Ariz. Rev. Stat. Ann. §§13-701, 13-801, 23-4128, 23-418.01, and 13-803 (Supp. 1986).
12. For example, under Kentucky's state plan regulations for controlling hazardous energy (i.e., lockout/tagout), locks would be required rather than locks or tags being optional as under the federal standard. Lockout/tagout is discussed in more detail in Chapter 2.

13. 29 U.S.C. §667.
14. 29 U.S.C.A. §667; 29 C.F.R. §1902.
15. The states and territories operating their own OSHA programs are Alaska, Arizona, California, Hawaii, Indiana, Iowa, Kentucky, Maryland, Michigan, Minnesota, Nevada, New Mexico, North Carolina partial federal OSHA enforcement, Oregon, Puerto Rico, South Carolina, Tennessee, Utah, Vermont, Virginia, Virgin Islands, Washington, and Wyoming.
16. 29 U.S.C. §655(b).
17. 29 U.S.C.A. §655(b)(5).
18. 29 U.S.C. 1910.
19. 29 C.F.R. §1911.15. (By regulation, the Secretary of Labor has prescribed more detailed procedures than the OSH Act specifies to ensure participation in the process of setting new standards, 29 C.F.R. §1911.15.)
20. 29 U.S.C. §1910.
21. 29 U.S.C. §655(b).
22. 5 U.S.C. §553.
23. 29 C.F.R. §1911.15.
24. *Taylor Diving & Salvage Co. v. Department of Labor*, 599 F.2d 622 7 O.S.H. Cases 1507 (5th Cir. 1979).
25. 29 U.S.C. §655(c).
26. Ibid at §655(b)(1).
27. Ibid at §656(a)(1). NACOSH was created by the OSH Act to "advise, consult with, and make recommendations ... on matters relating to the administration of the Act." Normally, for new standards, the Secretary has established continuing committees and ad hoc committees to provide advice regarding particular problems or proposed standards.
28. *OSHA Compliance Field Operations Manual (OSHA Manual)* at XI-C3c (Apr. 1977).
29. Ibid.
30. Ibid at XI-C3c(2).
31. Ibid at (3)(a).
32. 29 U.S.C. §666.
33. Ibid at §666(b).
34. For example, if a company possesses 25 identical machines, and each of these machines is found to have the identical serious violation, this would theoretically constitute 25 violations rather than one violation on

25 machines, and a possible monetary fine of $175,000 rather than a maximum of $7,000.

35. *Occupational Safety & Health Reporter*, V. 23, No. 32, Jan. 12, 1994.
36. *See Infra* at §1.140.
37. 29 U.S.C. §666(e)–(g). *See also OSHA Manual*, supra note 62 at VI-B.
38. A repeat criminal conviction for a willful violation causing an employee death doubles the possible criminal penalties.
39. 29 C.F.R. §5.01(6).
40. *U.S. v. Dye Const. Co.*, 510 F. 2d 78, 2 O.S.H. Cases 1510 (10th Cir. 1975).
41. *U.S. v. Crosby & Overton*, No. CR-74-1832-F (S.D. Cal. Feb. 24, 1975.)
42. 18 U.S.C. §2.
43. *See L.A. Law: Prosecuting Workplace Killers*, A.B.A. #J. 48 (Los Angeles prosecutor's "roll-out" program could serve as model for OSHA.)
44. *See infra.*
45. *See infra* §§1.10 and 1.12.
46. 29 C.F.R. §1903.8.
47. 29 U.S.C. §17(f). The penalty for providing advance notice, upon conviction, is a fine of not more than $1000 or imprisonment for not more than 6 months, or both.
48. *Occupational Safety and Health Law*, 208–09 (1988).
49. 29 C.F.R. §1903.7(b) [revised by 47 *Federal Register* 5548 (1982)].
50. *See, for example,* 29 C.F.R. §1903.9. Under §15 of the OSH Act, all information gathered or revealed during an inspection or proceeding that may reveal a trade secret as specified under 18 U.S.C. §1905 must be considered confidential, and breach of that confidentiality is punishable by a fine of not more than $1000 or imprisonment of not more than 1 year, or both, as well as removal from office or employment with OSHA.
51. It is highly recommended by the authors that a company representative accompany the OSHA inspection during the walk-through inspection.
52. *OSHA Manual, supra* note 62, at III-D8.
53. 29 C.F.R. §1903.7(e).
54. *OSHA Manual*, supra at n. 62, at III-D9.
55. 29 U.S.C. §660(c)(1).
56. Ibid at §658.
57. *Fed. R. Civ. P.* 4(d)(3).
58. 29 U.S.C. §659(a).

59. *Preparing for an OSHA Inspection*, Schneid, T., Kentucky Manufacturer, February 1992.
60. 29 U.S.C. §651 et seq., §9(a) (1970).
61. Ibid. The statute of limitations contained in §9(c) will not be vacated on reasonable promptness grounds unless the employer was prejudiced by the delay.
62. Ibid.
63. Ibid.
64. 29 U.S.C. §651 et seq., §10(a) (1970).
65. *B.J. Hughes*, 7 O.S.H. Cases 1471 (1979).
66. 29 U.S.C. §651 et seq., §10(1970).
67. M. Rothstein, *Occupational Safety and Health Law*, (2d ed. 1983), summarized and reprinted in *Employment Law*, Rothstein, Knapp, and Liebman, Foundation Press (1987) at 509.
68. *See, for example*, *Atlas Roofing Co. v. OSHRC*, 430 U.S. 442, 5 O.S.H. Cases 1105 (1977). The Supreme Court also held that there is no Seventh Amendment right to a jury trial in OSHA cases.
69. Rothstein, Knapp, and Liebman, *Employment Law*, The Foundation Press (1987) at 599.
70. 29 U.S.C. §651 et seq., §11(a) (1970).
71. 29 U.S.C. §651 et seq., §11(b) (1970).
72. *Gilles v. Cotting*, 3 O.S.H. Cases 2002, 1975–76, OSHD §20,448 (1976).

CHAPTER 11

CAPTURING THE ENTREPRENEURIAL SPIRIT

> Treasure your entrepreneurs, because it is from their unfettered and sometimes undisciplined efforts that job creation will come.
>
> —David Birch

> I am first and foremost a catalyst. I bring people and situations together.
>
> —Armand Hammer

Employees in the past often perceived their jobs as simply punching the time clock, doing what the boss told them, and picking up a pay check. Today's employees expect far more than their forefathers. The employees of today may be shareholders in the company; prize individualism; expect, if not demand, to be treated appropriately by management; work with their mind instead of their back; will express their needs and wants; and will not hesitate to move to another company if their the needs are not being fulfilled. However, safety and resource control professionals often continue to use the tried-and-true methods of managing this important function, foregoing the potential of uncovering new and better methods of achieving our ultimate goals and objectives.

Safety and resource control professionals should look "outside the box" for new and creative methods in which to involve the employee, their families, the community, and other professionals within the scope and breadth of their safety and resource control efforts. There is no downside to involving as many different individuals or groups as possible in order to motivate your employees to embrace safety and resource control and make safety and resource control part and parcel of their daily work activities. We must strive to ensure that each and every employee makes safety and resource control as essential and as routine as putting on their personal protective equipment before beginning their shift. Safety and resource control is a "mind set" and we must locate creative methods of maintaining safety and resource control in the front of each and every employee's mind each day. Through this repetition, safety and resource control can become ingrained in their thoughts to the point that the ideas take hold, the employee becomes self-motivated, and new and creative ideas are created to accelerate the safety and resource control program to greater heights.

Safety and resource control professionals should be searching for new and creative ideas on a daily basis. There usually is no blueprint to be followed, which is why safety and resource control professionals must be diligent and observant for new ideas, listen actively to employees, and search for new and novel methods of energizing their program.

The safety and resource control professional of today faces new and complex problems and issues which have not been previously addressed by their predecessors. Often, there is no standard addressing the particular issue and there is no framework or road map to follow. With these new issues, safety and resource control professionals should search beyond the boundaries of traditional safety and resource control and use their creativity and ingenuity to analyze and identify other solutions to address these issues.

A major issue facing many employers that has a direct impact on the safety and resource control efforts is the selection and hiring of qualified employees. Many safety and resource control professionals have established exceptional programs, but their injury and illness rates continue to escalate, with the cause of this escalation blamed on the lack of qualified candidates. In the tight labor market of today, more than one safety and resource control professional has been heard blaming their increased injury and illness rate on the bottom-of-the-barrel employees who were hired because "we needed warm bodies for production."

Once the individual is hired, the responsibility for the safety of the employee in the workplace belongs to the company, and thus the safety and resource control professional. Complaints from safety and resource control professionals range from the employees' drug or alcohol usage to the new employees' inability to be educated. However, if the employees injure themselves on the job, they often become long-term wards of the safety and resource control professional under the dual workers' compensation responsibilities. Safety and resource control professionals often feel that they are captured in a continuous vicious circle.

To combat these issues, many safety and resource control professionals have reevaluated the entire selection, hiring, and training process in an effort to pinpoint potential problem areas and individuals before hiring and have instituted new methods of evaluating candidates and training employees. Some of these creative ideas include the following:

- Look outside the usual employee candidate pool. Many employers simply run an advertisement in the newspaper and take the best of the individuals who show up at the door. To widen the application pool, some employers have targeted nontraditional candidate groups such as retired individuals, special education groups, disabled groups, and school-to-work programs to acquire qualified candidates. In most circumstances, the candidates within the nontraditional groups possess an exceptional work ethic and lower incidents of injury and are generally good employees. Safety and resource control professionals often voice a concern with regard to compliance with such laws as the Americans With Disabilities Act and related laws for individuals with disabilities. In most circumstances, if a reasonable accommodation is necessary, the accommodation can be made at a nominal cost. Several sources of financial and accommodation assistance are available to help the safety and resource control professional in providing such accommodation.
- Approximately 50% of all employers in the United States use some form of alcohol and controlled substance testing to identify individuals using drugs and alcohol within the pre-employment or post-employment offer stage of the employment process. Most safety and resource control professionals have found that this

type of testing has provided a reduction in their incidents of work-related injuries and illnesses.

- Some employers use various forms of psychological testing, such as the MMPI, as well as physical testing, such as physical agility testing, within the post-offer stage of the selection process. These types of tests can often identify restrictions for specific individuals and often help in the proper placement of employees.
- Most employers use a new employee orientation/training program of various types to indoctrinate newly hired employees as to the expectations of the job. Although some employers provide only a "bare bones" session consisting of completing tax forms and reading the plant rules, many safety and resource control professionals have found that new employee orientation is an exceptional opportunity to start new employees off on the right foot in safety and resource control. A new employee orientation can be utilized to train individuals as to safety and health rules and regulations, location of programs, proper wearing of personal protective equipment, and numerous other aspects of safety and resource control.

In the area of engineering controls, safety and resource control professionals are encourages to look outside the traditional area of basic machine guarding to identify new technologies for adaptation within the safety and resource control area. Just as computers have revolutionized business in general, computers have been adapted to numerous uses in the safety and loss prevention area from cataloging material safety data sheets to tracking accident reports. Are there other uses for the computer to assist safety and resource control professionals in their jobs?

New technologies and products with potential uses within the safety and resource control area have been developed and continue to be developed virtually on a daily basis. For example, machine guarding twenty years ago primarily consisted of metal cages over moving parts or cables to remove the hands/legs from the contact point. Today, technology have provided such guarding as light curtails and laser guided operations. This new technology was "science fiction" not long ago; however, it is available over-the-counter today. Are there other new technologies that can be adapted to the safety and resource control area?

In the area of employee motivation and behavior modification, do individual employees possess the same "motivating triggers" as employees in the past? Does the employee who grew up in the 1960s possess the same motivations as an employee who grew up in the 1980s? Are the expectations of employees today different from those of their fathers or mothers? Is the education provided today to children in our school systems different from the education most safety and loss prevention professionals received during their formative years? How many of today's safety and resource control professionals spent time in day care during their early years? How many of today's safety and resource control professionals worked during their high school years? How many of today's safety and resource control professionals had a television, let alone a personal computer?

For most safety and resource control professionals, there is a significant difference between the background and motivations of employees in the past and today's employee. Then why are safety and resource control professionals still using the same old methods of attempting to motivate employees to work safely? Do they think that if the "carrot and the stick" approach worked in the past, it will work in the future? Is the "carrot" different today? Is the "stick" different?

Safety and resource control professionals should search for the new triggers within their employee population which will motivate them to work in a safe manner. Traditionally, safety and resource control professionals used safety incentive programs such as green stamps for not getting hurt for a period of time or monetary bonuses for reaching a specified goal to reduce lost time days. Most safety incentive programs were short lived and focused on traditional Occupational Safety and Health Administration (OSHA) recordkeeping requirements. The safety and resource control incentive motivated employees during the period of the "contest," but the overall safety and resource control efforts often suffered upon achievement of or loss thereof the monetary or physical incentive. Traditional safety incentive programs often motivated employees to hide injuries or illnesses until after the contest period and often masked deficiencies in the overall safety and resource control programs.

Although the author is not a proponent of safety and resource control incentive programs until all foundational components of the program are in place and functioning properly, are there other motivational triggers that can be used to increase employee awareness?

Safety and resource control professionals may wish to consider the following ideas:

- A simple pat on the back and telling employees they are doing a good job.
- "I saw you doing something right" cards.
- Safety-related birthday cards.
- Congratulations on bulletin boards.
- Saying "Thank you" in employee newsletter.
- New ideas contests.
- Safety information provided to the employee's family.

These type of incentives do not cost anything in terms of monetary expenditures however they are great motivators for individuals. Little things mean a lot. Individual praise or recognition goes a long way in motivating individual employees. Positive motivation has been found to be a greater motivating factor than negative reinforcement. When is the last time you told your employees they were going a good job?

In lieu of the traditional safety and resource control incentive programs, some professionals have tested new and innovative methods of motivating their employees such as the following:

- The winning safety group gets to run out on the field and be recognized at a local college or high school football or basketball game.
- The winning individuals get to shoot for a large prize at intermission in front of the crowd at a local semi-pro hockey game.
- A clown with balloons and pizza treats the group at the plant.
- The winning individuals get to putt at a local golf course for prizes.
- The "bosses" are required to perform the employee's job for a day, and the employee gets the day off.
- The safety and resource control professional has to kiss a pig.
- The winning group is taken roller skating.

These types of safety and resource control incentive programs are new and different for most employees and will definitely be the center of discussion around the water cooler for a number of weeks. Most employees with any type of longevity in the industrial workforce has participated in the traditional programs. Safety and resource control professionals are encouraged to talk with their employees and find out what "trips their trigger" or motivates them. The employees will tell the safety and resource control professional their likes and dislikes, and the safety and resource control professional can design a creative program to trip their motivational trigger. Remember, there is no law prohibiting fun.

I am not a proponent of the multitude of theories and canned programs regarding employee behavior modification until the foundational elements of the safety and resource control program are in place. As with safety and resource control incentive programs, the vast majority of these programs will be short-lived and not cost effective if the basic foundational elements of the safety and resource control program are not in place and functioning properly. However, for those programs with a solid foundation looking to push the program to a higher level, safety and resource control professionals may consider incorporating one or more of the behavioral modification programs into the mix.

Safety and resource control professionals are urged to look beyond the traditional areas for new and creative way of involving and motivating employees and management in the safety and resource control efforts. As Aristotle said many centuries age, "All things grow old through time." Many safety and resource control programs have grown old with time and have often lost their effectiveness. Safety and resource control professionals should look beyond the norm for new and innovative ways to energize and constantly improve your programs through exploration on the pathways are in areas less traveled. And, above all, remember to make the activities fun!

CHAPTER 12

THE TEAM IS BETTER THAN THE SUM OF ITS PARTS

> Coming together is a beginning; keeping together is progress; working together is success.
>
> —Henry Ford, Sr.

> A major reason capable people fail to advance is that they don't work well with their colleagues.
>
> —Lee Iacocca

To be successful in the safety and resource control function, the safety and resource control professional must be the leader, the guide, the shepherd, the coach to steer the entire team consistently toward the goal line. The safety and resource control professional cannot be a one-person band but must be an integral component, a driving force, in guiding the team to ultimate success. Safety and resource control is everyone's job on the team and the safety and resource control professional should provide the tools, guidance, or even the "kick in the pants" to maintain safety and resource control at the top of the priority list.

What is often heard in the industrial environment is: "Safety and resource control is not my job. It's the safety person's job." In this type of environment, employees have not achieved a level of understanding that safety and resource control is their job, not the safety and resource control professional's job. It is the safety and resource control professionals job to plan, organize, direct, control, energize, educate, motivate, and otherwise guide the safety and resource control efforts, but it is the individual team member's responsibility to integrate safety and resource control into their daily work activities in order for the organization to achieve the safety grail (see Chapter 4). It is the safety and resource control professionals responsibility to ensure that all team members know their responsibilities, are provided the tools and held accountable for the achievement of their identified objectives.

Safety and resource control professionals who often fail, burn out, or become frustrated in the job are most probably "doing" rather than managing. Given the numerous hats that a safety and resource control professional wears on a daily basis, if the safety and resource control professional is performing a job which can or should be being performed by another team member, another aspect of the program is falling down or deteriorating. It is often a major juggling act for many safety and resource control professionals to "keep all the balls in the air" on a daily basis. The problem arises, however, when a safety and resource control professional drops a ball, someone gets injured, something gets broken, or something (usually expensive) bad goes wrong! Safety and resource control professionals should manage the function just as production, quality, or other functions within the organization are managed.

Safety and resource control professionals should search and locate a management style that best fits their situation, personality, and corporate culture. The basic components of virtually all management styles involve the components of planning activities and programs, organizing these activities and programs, directing the activities and programs, and controlling them. The management styles and techniques may vary, but it simply boils down to the fact that the safety and resource control function must be managed as with any other function within an organization.

There are usually several areas within the management of the safety and resource control function that often creates difficulties for the safety and resource control professional in directing and managing the team efforts. First, and almost universally heard, is the issue of time management. As we are all aware, a safety and resource control professional does not work the traditional 9 to 5. There is always more to do than hours in the day. However, safety and resource control professionals should learn to prioritize their activities and effectively manage the precious minutes and hours in any given day in order to maximize their efforts. The light at the end of the tunnel is the fact that the better safety and resource control professionals become at effectively managing time and programs, the more time they will have to interact with team members and actively promote the safety and resource control function. If essence, it's difficult to be effective when all of your time is allotted to putting out fires.

Another area that often eats up the time for safety and resource control professionals it team member problems. The general rule is safety and resource control professionals will spend 90% of their time with 10% of the team members and 10% of their time with the other 90% of the team members. Safety and resource control professionals are often confronted with is the team member who continues to exhibit unacceptable behaviors despite your positive reenforcement and negative reenforcement. These are often the weakest links in your overall safety and resource control efforts. Through your coaching or counseling sessions, the safety and resource control professional may be able to identify the underlying reasons through which, or for which, the team has continued to exhibit the unacceptable behaviors. However, in some circumstances, evidence of positive reenforcement or early negative reenforcement does not change the offenders unacceptable behaviors.

To this end, the safety and resource control professional should be able to identify the underlying reason that the employee will not or cannot change the unacceptable behavior. Many of the reasons for which the behavior is not being changed could include:

- The employee is simply working the job in anticipation of a better job.
- The employee is experiencing family problems, drug or alcohol problems, or other outside influences.
- In an attempt to be part of the team, the employee continues the unacceptable behavior, to show off for the other employees.
- The employee's unacceptable behavior is a method through which to rebel against management.

Given the time, effort, and expense to train and prepare the employee to perform the job function, it is important that the safety and resource control professional discover or uncover the source of the constant offender employee's problem areas. The problem areas can be identified, appropriate assistance, such as Employee Assistance Programs (EAPs), can be provided for the employee if available. If the problem area cannot be discovered and the employee is unwilling to work with the safety and resource control professional, supervisor or team leader, the ultimate result normally is involuntary termination from employment at some point in time. Remember, documentation of your attempts to assist the constant offender will be scrutinized in the future. Careful documentation throughout the entire process is essential.

Disgruntled employees, as with the constant offender, can require a substantial amount of your time due to the sheer number of sessions in an attempt to modify their unwanted behaviors. Disgruntled employees, unlike the constant offender, can often possess an exceptional attendance record, work performance record, and be an exceptional employee if they did not constantly complain about virtually everything.

Disgruntled employees or the constant complainer underlying reasons for their inappropriate behavior, which can include the following:

- Loss of control—the disgruntled employee may be in a job in which he or she has little control.
- The disgruntled employee needs to voice their opinion.
- The disgruntled employee is getting no response or feedback with regard to complaints.
- The disgruntled employee may be serving as the unofficial leader for their work area.
- The individual may be having personal difficulties outside of work that are being vented through their actions on the job.

Safety and resource control professionals should attempt to identify the underlying reason for which the employee is disgruntled or constantly complaining. Often, involvement with company activities, such as the safety committee, can permit the employee a method through which to voice his/her concerns in the appropriate manner. Safety and resource control professionals should not completely discredit the complaints coming from a constant complainer given the fact that the allegedly disgruntled employee may be the unofficial leader of a group of employees who are experiencing problems in their work area. Safety and resource control professionals should strive to turn what appears to be a negative behavior into a positive behavior for the employee through appropriate means within their organization or company.

There will be numerous constraints on the safety and resource control professionals time and abilities. The safety and resource control professional must effectively manage his/her time as well as programs, team members and other essential functions in order to achieve success. However, it is easy to lose focus on the true direction when your up to your backside in alligators. Manage your self, your time, and your programs and strategically use your most valuable resource—yourself—to search for the weakest links in your program to make appropriate corrective measures. Remember, you are only as strong as your weakest link.

CHAPTER 13

GIVE 'EM THE TOOLS TO BE SUCCESSFUL

> Excellence is an art won by training and habituation. We are what we repeatedly do. Excellence, then, is not an act but a habit.
>
> —Aristotle

> Some men make difficulties, and some difficulties make some men.
>
> —Anonymous

It is imperative that safety and resource control professionals provide each individual with the tools for success. Within the safety and resource control function, this not only means personal protective equipment and machine guarding, but the skills and abilities through training and education to perform the job safely and effectively on a consistent basis.

To this end, safety and resource control professionals are often confronted with a wide variety of different types of training and education programs for their management team and employees. In most organizations or companies, the range of training can vary from new employee orientation training through education of the board of directors in specific topic areas. Safety and resource control professionals should always prepare properly before initiating any training and ensure that all materials, curriculum, audiovisual equipment, and every other aspect of the training is choreographed to ensure

maximum efficiency and complete understanding by the receiving parties.

One of most important types of training conducted in most organizations and companies is that of a required government agency Compliance Training and Education Program. This type of training possesses very specific program elements that must be meet, as well as documentation requirements, in order to prove that this training has been conducted in accordance with requirements. Compliance training can include, but not be limited to, training for specific occupational safety and health standards (e.g., lockout/tagout, hazard communication), environmental protection agency training, sexual harassment training, and a myriad of other required training and education programs.

Many organizations offer training and education programs to inform, as well as motivate, employees and management team members. In many cases, this training may provide specific methodologies as to improvement of management skills and other basic skills. Motivational training is often incorporated into a series of training programs at various levels designed to improve performance, communications skills, and other areas.

In many organizations, new employees and or employees changing departments or areas are required to go to participant in a new employee orientation and training program which encompasses a large variety of topic areas ranging from where to acquire your paycheck to the safety and resource control rules. This training is crucial, as it is the first chance the safety and resource control professional has to interact with the new employees and set the tone for the employees future with your company.

With virtually every type of training and education program, some type of follow-up evaluation and monitoring is required in order to ensure that the information transmitted during the training program is being used by the employees or management team member. This training can also include compliance refresher training as required by some requirements as well as specific follow-up reenforcement training for management skills.

In training and education activities, the safety and resource control professional may be faced with a wide variety of educational levels through which to transfer important information. The safety and resource control professional must identify the educational level of the participants in the specific training program to adjust the materials and program to transmit the information to the participants at the appropriate level.

Don't forget that your upper management group must also be educated in the safety and resource control function and activities. In many organizations, the safety and resource control professional is responsible for training and educating the upper management group, including, but not limited to, board of directors members, presidents, chief executive officers (CEOs), plant managers, and others in various specific topic areas. These topic areas may include modification in the laws, recent court decisions, new Occupational Safety and Health Administration (OSHA) standards, new Environmental Protection Agency (EPA) requirements, and other specific topic areas. The education level in which most education and training programs offered at the upper management level is that of an advanced learner or college level.

At the advanced level, training and education are normally focused on a specific topic and the information provided is very focused. Supplemental materials in this type of training can normally include full case studies, advance research information, full text of the law, and other unedited documents. Often with this type of training, a summary of the information should be provided and provided to these busy participants as well as full text materials. The use of audiovisual materials is normally limited to very focused items, and the training time provided is normally of shortened duration.

Safety and resource control professionals are often responsible for the development of education and training programs for their supervisors, team leaders, and other members of mid-level management. This training is often geared to the high school level; however, some organizations increase the educational level through college level, depending on the topic area and background of these supervisors or team leaders. This training can include a variety of areas from management skills training through specific regulatory compliance programs.

In many organizations, the safety and resource control professional is responsible for the development of new employee orientation programs as well as ongoing employee training, which is normally focused primarily on the area of compliance. Employee training is normally developed at approximately an eighth to tenth grade level dependent on the background of the work force.

The training should be designed not only in accordance with the educational level of the participants, but also in different formats to maximize the level of understanding. With the new technologies, training can now be provided by interactive computer programs or can be transmitted to the location by distance learning. Several formats for consideration include the following:

- *Classroom (formal) training:* This is the traditional method of learning with an instructor providing information to the participants. This format can be provided in lecture and/or facilitating manner.
- *Hands-on training:* Especially in the area of compliance or job-specific training, the training program can be focused on the employee actually performing the required functions under the supervision of a trained instructor or supervisor.
- *Computer-assisted training:* Depending on the level of technology available in the facility, computer-assisted training using compressed videotape or designed programs, or both, can be provided to employees.
- *Interactive computer training:* The newest technology permits the participant to interact with the computer program to acquire the necessary education level in testing within the software program.

The goal of virtually every level of training is to communicate properly the required information to the participant and ensure complete understanding of the information. Depending on the organization in the type of resources available, the safety and resource control professional can adapt the information and develop a training program to meet the needs of the specific participants.

From the time we were born, we have acquired information using our senses of sight, speech, smell, hearing, and touch. In the industrial world, the primary senses we use to learn are sight, speech, hearing, and touch, with the primary senses sight and hearing.

Traditional learning was focused primarily on the senses of sight and hearing as well. However, safety and resource control professionals are encouraged to explore additional avenues through which to increase the level of understanding as a particular participant and focus the training around the specific subject matter. We have become accustomed to watching a television or using a computer screen to acquire information. Most adult learners acquire information primarily through sight and verbal explanation. Adult learners have a greater attention span, but they require an entertainment factor to maintain a high level of concentration.

With specific technical skills, incorporating the use of touch can increase the level of understanding. As with riding a bicycle, we can easily tell someone how to ride a bike and show an individual how to ride a bike but until they actually get on the bicycle, the proficiency's of their skills in riding a bicycle will not be maximized.

- *Visual learning:* Safety and resource control professionals can maximize the visual learning capabilities of adult learners through the use of videotapes, slides, overheads, and computer-generated information.
- *Auditory learning:* Verbal information must be presented in a decibel level range (not too high too low) and in an entertaining manner to maintain the interest of an adult learner. Monotone voices, screechy voices, noise interference, and other distractions can affect auditory learning.
- *Learning through doing:* Hands-on training has been found to be the most effective method, when combined with other types of training, to acquire proficient skills and employees. This additional use of the employees sense of touch increases the learning curve.

The time provided by most organizations for education and training programs is normally limited. Safety and resource control professionals are encouraged to maximize this limited time in order to achieve the most effective results. To do this, the safety and resource control professional must plan and organize the training activity effectively and present the information in the most effective manner.

Safety and resource control professionals should appropriately schedule training and education activities to maximize the learning of the employees or management team members during this limited period. To do this, the safety and resource control professionals should effectively plan every aspect of the activity from the setting arrangement to the final testing. All curriculum, audiovisual aids, documents, supplemental material, and other required items must be prepared in advance to maximize efficiency.

In most organizations, the time provided for education and training programs is normally before the shift starts, immediately after the shift, at lunchtime, or other nonproductive work hours. These are normally not the best time periods for learning for most individuals. Before work, the individuals are in a routine and are not normally fully prepared to learn. After a work shift, the employee may be tired after a full day's work and may be unable to focus properly on the information being provided. During lunchtime and breaks, employees are often busy eating lunch or taking care of other normal activities while the information is being provided. To maximize the participant's learning, the safety and resource control professional should attempt to schedule training and education programs during the maximum peak period for the individual. This will vary depending on the number of shifts (i.e., 3–11 PM; 11–7 AM) and the safety and resource control professional should be flexible in scheduling training to achieve maximum results.

Every minute of a training program is crucial. The maximum amount of information in which the individual can absorb and retain should be incorporated into the training agenda. To do this, the safety and resource control professional should properly schedule and design an agenda to be closely followed during the training program. This does not mean that time should not be allotted for employees to ask questions or acquire clarification of information; however, training programs are not the time to voice complaints or discuss non-issue-related items.

Preparation of training materials can include preparation of audiovisual materials, written handout materials, and training manuals. In some circumstances, the safety and resource control professional may be able to acquire a "canned" program that is readily available on the market. In most circumstances, the safety and resource control professional should create this information and prepare the information for presentation to the participants, the employees, or management team members. To do this, the safety and resource control professional should have access to the appropriate research materials, which often include OSHA, EPA, and other regulations. Sources that may be used by the participant or by the safety and resource control professional can include implant sources, local libraries, universities, or most recently the Internet. Especially in the area of compliance, a substantial number of the government regulations and supplemental information can be acquired through an Internet source at your facility.

The level in which the educational materials are presented to the employees or management team members is essential. As a general rule, educational materials should be designed for the lowest level of education within the specific group of employees or management team members. The employees and management team members at a higher educational level can understand the information at a lower level however the employees at a lower may not be able to understand the higher level of information. And above all, keep it simple.

Is it better to give out your materials before the training session? Is it better to hand out your materials during the training session? Is it better to give the materials after the training session? This would depend on the organization. Giving materials out before the training session often provides the employer management team member a chance to review the materials before training; however, it has been found that often the employee forgets the training materials or fails to review this information. Handing out written information during a training session will direct the employee or management team members attention to the written material rather than the spoken communications. If materials are being handed out during the training session, it is often better to allow for a period of time during which the employee or management team can read the information before providing additional oral communication. Handing out written information after the training session gives employer management team members the opportunity to refresh their memory after completion of the course and for use of reference materials. However, in many organizations, written material gets deposited in the waste paper basket immediately after the oral presentation.

The acquisition of feedback and follow-up evaluation after completion of the training session is essential. Employees should be given an opportunity to provide feedback during the training session; however, it is often important to ask for their thoughts and ideas after the training session. Follow-up and reenforcement of the ideas are crucial to the learning process and should be included in any training program.

One method in which safety and resource control professionals can enhance the overall learning of the employees and management team members is through the use of audiovisual aids. Audiovisual aids should be used only to supplement the training program and not for training in and of itself. Audiovisual aids can include, but not be limited to, transparencies, slides, videotape, computer animation, computer-generated programs, and other audiovisual aid formats.

Safety and resource control professionals will find that there is a wide variety at varying cost levels available "canned" programs. These programs are normally generic in nature and often need to be modified for plant-specific activities and operations. Good sources to look for such programs is the published safety directories, professional magazines, and libraries.

With today's technology, a number of disc and CD-ROM programs are readily available in a generic format for use on a wide variety of topics. Specific software programs can be developed for use by plant employees and management team members in a wide variety of formats. Computer-generated and CD-ROM programs are normally listed in the professional magazines and computer publications.

Although the initial cost tends to be high, a number of interactive video and interactive training programs are currently on the market. These programs tend to be generic in nature, but they do provide the opportunity for employees and management team members to work through the program at their own pace on a personal computer.

In virtually all compliance areas as well as many management skills areas, generic "canned" written materials are available in a wide variety of educational levels ranging from the comic book version through full text. This information is normally copyrighted and thus a site permit or per piece cost would be required. This information is normally generic in nature and is often difficult to modify for individual plant situations.

Depending on time constraints, budgetary constraints, and the training topic, the safety and resource control professional may want to consider the wide variety of audiovisual aids and supplemental materials that are available on the market. Conversely, specific topic areas over budgetary constraints may limit the available safety and resource control professionals may want to consider the development of materials in-house and the use of a VCR to develop in-house videotapes. Above all, the safety and resource control professionals should use the available resources in an efficient manner and maximize the learning capabilities of the employees or management team members. Above all, remember that simply putting a videotape in a VCR is not training. Use every possible source to maximize the efficiency of your training.

Preparation is the key to success in any type of training and education program for employees or team members. Safety and resource control professionals should properly prepare the training materials including training schedules, curriculum, supplemental materials, and preparation for any training session. The training materials must be focused on the participants' educational level and the specific training topic. The materials must be well written and presented in a manner in which the employee or management team member can absorb the information within a reasonable period of time.

In many circumstances, the safety and resource control professional cannot go on the open market and acquire a training program focused specifically on the needs of the individual employee or management team member. Thus, the safety and resource control professional should prepare the training materials, course curriculum, supplemental materials, and other information in preparation for presenting the training program. Using the "no stone unturned" theory, the safety and resource control professional should prepare the training program to encompass every facet from the preparation through the follow-up training to ensure the most efficient method of transmitting this important information.

Safety and resource control professionals should be aware that training information can be acquired from numerous sources. This can include, but not be limited to, existing training manuals, videotape training courses, government publications, OSHA standards, and other written sources. Safety and resource control professionals should inquire as to the availability of education and training assistance through local universities, state plan programs, and federal and state agencies.

Safety and resource control professionals should also explore possible sources of outside funding to assist in your training and education endeavor efforts. Often, federal, state, and even local programs will provide monetary assistance to companies and organizations for specific training and education purposes.

With most training materials, less is better. The educational level should correlate to the specific employee or management team member, and the lowest educational level should be used. In many circumstances, employees do not have the time to read a large volume of written material. They want a short course or summary. Information presented in training may be in an outline form or bullet point form, focusing on the highlights for immediate review and understanding by the employee or management team member.

There is no substitute for properly preparing your employees at all levels with the skills and abilities to be successful and safe in their jobs. Training and education are important facets of every safety and resource control professionals job function and should not be provided second fiddle position on the priority list. Training and education should be address with the same intensity as every other component of the safety and resource control function. Remember, if we can plant the seed of safety and resource control in their mind, we will no doubt reap the harvest down the road.

CHAPTER 14

THINKING OUTSIDE THE TRADITIONAL BOX

> Let us think of quietly enlarging our stock of true and fresh ideas, and not, as soon as we get an idea of half an idea, be running out with it into the street, and trying to make it rule there. Our ideas will, in the end, shape the world all the better for maturing a little.
>
> —Matthew Arnold

> Your most brilliant ideas come in a flash, but the flash comes only after a lot of hard work. Nobody gets a big idea when he is not relaxed and nobody gets a big idea when he is relaxed all of the time.
>
> —Edward Blakeslee

With the world of safety and resource control continually changing, safety and resource control professionals are encouraged to look beyond the book and the tried and true for new and creative ways of improving the overall function. Safety and resource control professionals should search for new ideas, new programs, and new methods through which to achieve the continuous growth and development of their function. Safety and resource control professionals work within the "box" created by regulatory requirements, management constraints, time, and financial constraints, but new ideas and new concepts should not automatically be disgarded simply because the idea or concept has not been tried elsewhere.

The standard joke in the safety and resource control profession is that usually research and development (R&D) means rip-off and duplicate. Safety and resource control professionals often find a program that has been successful elsewhere and attempt to duplicate the success in their facilities. This is not to say that there is not a time and a place for this methodology but many safety and resource control professionals have become locked in this methodology and have foregone the possibility of trying anything which does not possess a substantial track record.

Safety and resource control professionals are encouraged to look "outside of the box" for new and creative ways of improving not only the in-plant safety and resource control function but to the overall quality of the lives of their employees, community, and the profession. These ideas cannot be found in the Code of Federal Regulations. New and creative ideas should be analyzed and, if feasible, attempted. Most individuals learn more from their failures than from their successes. If the idea doesn't work the first time, correct the deficiencies and try again. If the safety and resource control professional does not try new ideas and methods, nothing will be gained except the status quo.

Creative opportunities abound in the area of community involvement, especially in the area of safety and resource control. Virtually everything that is done possesses a safety and resource control component. All that is left is for the safety and resource control professional to become involved. Safety and resource control professionals, mainly because of their job function, are often the visible manifestation of the company within the community. The interaction between the company and the community usually involves safety and resource control issues (e.g., community disaster preparedness planning) that reflect on the efficacy and reputation of the company.

The image and reputation of the company as often viewed through the safety and resource control professional, have a direct and indirect bearing on a number of important issues for your company. For example, on a positive note, if your company is perceived as a good corporate citizen, individuals seeking employment will be referred to your company by word of mouth, increasing your applicant pool of qualified potential employees. Conversely, if your company is not perceived as a good corporate citizen, your company would have a hard time attracting qualified employees and thus may be required to pay a higher wage to attract employees.

The interaction with the community is also important in peripheral matters such as with interaction with the local medical community. A company that has a positive interaction with the medical community creates a bond of trust and cooperation. Thus, if an employee is injured and the safety and resource control professional works with the medical community and the employee, the workers' compensation costs may be able to be reduced through returning the injured employee to restricted or light duty.

If the image and reputation of the company are poor, the medical community may want to maintain the injured employee off work, thus increasing the workers' compensation costs because of the fear of re-injury or failure to follow the proscribed medical restrictions.

Safety and resource control professionals should search for ways in which to involve their company in community activities and become or maintain the good corporate citizen status. Generally, people think the worst of the things they do not know. What do the citizens who live around your operations think happens at your facility? If all they see is ambulances pulling in your plant everyday, what perception will they have of your safety program, and therefore your company? Is the exterior of your facility littered with debris, or is it manicured grass? Does this affect your company image and what the citizens think about your company?

How can safety and resource control professionals have an impact on the image and reputation of company in the community? The easiest and simplest explanation is to become involved in community activities. The job responsibilities in safety and resource control require involvement with community officials through such regulations as the area of disaster preparedness and the community's right to know. Does your operation's preparedness planning include involvement with the local fire department, law enforcement, and medical community? Have the community officials been provided a plant tour so that they know what to expect in an emergency situation? Is there a coordination of activities in preparing for an emergency situation, such as correlating communications systems? Do local officials know how much funding is provided to the community through corporate, employment, and other taxes?

Safety and resource control professionals can also get involved in the community through participation in organizations such as the local Chamber of Commerce and related organizations. Safety and resource control professionals bring to the table a wealth of experience and expertise that is often needed in these types of volunteer organizations. This involvement not only assists the community but also creates an avenue of communication for safety and resource control professionals and their company.

Safety and loss prevention professionals can also lend their expertise in the area of community emergency preparedness committees, fire, and EMS committees and other local committees or groups. Safety and resource control professionals often take for granted the wealth of experience and ideas they possess that can enrich their communities.

Some of the other creative ideals that safety and loss prevention professionals may consider include:

- Plant tours for local schools.
- Speaking to local civic groups on safety.
- Serving on the local volunteer fire department or EMS.
- Sponsoring "Safety City" in your community.
- Sponsoring "Stop, Drop, and Roll" training at elementary schools.
- Serving on community "Stop Drug" committees and campaigns.

Safety and resource control professionals should search "outside of the box" for new ideas and methods to motivate their workforce. One of the main reasons, if not "the" reason why most employees get up in the morning and go to work every day is to support their family unit. The family unit is the underlying basis for the activities of most employees. The family is the "motivating factor" for many employees to do a good job and advance in the company. However, how often do we involve the family in the activities at work?

Most employees spend as many or more hours at work as with their family. Often, employees form a close bond with their co-workers at work or form a quasi-family unit. Does the employee's family know the person with whom mom or dad spends most of his/her time?

Moreover, do the members of the family even know why mom or dad go to work? Do they know what they do on a daily basis to earn a paycheck? Have they ever been to their workplace? Safety and resource control professionals often lose focus as to the underlying reason that people pull themselves out of bed in the morning and go to the workplace, day in and day out. The primary reason for most individuals is to support their family units.

Safety and resource control professionals should identify creative ways of involving the family in the work activities of mom or dad, especially in the area of safety and loss prevention. Can involvement of the family in the work activities of the employee create a greater bond in the company and the employee? Can the involvement of the family permit individual employees to think about what they are doing on a daily basis and to think about their own safety and the impact on their family if there is a safety breakdown? Can the family involvement in safety and loss prevention have an impact not only on the job but also at home?

Most employees are proud of their work. Safety and resource control professionals can provide opportunities through which employees can be rewarded by the people who are most important in their lives by providing opportunities for the family to become involved in the workplace. What ideas do you have to reward and motivate your employees?

Are new and creative ideas the sole domain of the safety and resource control professional? In order to tap into employee creativity, the safety and resource control professional should strive to create an environment in which employees are empowered to express their creativity and possess that freedom through which to express their creativity without any possible recourse. Who knows the equipment and operation better than the employees who work with the equipment and operations on a daily basis? But when is the last time we've asked for their ideas and input as to better ways to perform the job, perform the job in a safer manner, improve quality, or other aspects of the job?

Your employees are one of greatest sources of creativity and innovative ideas in the area of safety, health, environmental, and related areas if their creativity can be tapped. Enticing employees to express their creativity and ideas must be developed in conjunction with a strategic plan given the nature of the employer–employee relationship. Employees are often skeptical about expressing their wild ideas or about discussing blue sky concepts with management because of potential repercussions to their jobs or even that the "boss will think I'm dumb." An atmosphere should be created in which employees are encouraged to express their ideas no matter how far fetched.

To successfully acquire the vast storehouse of your employees creativity on an ongoing basis, a strategic plan in necessary and can be achieved through various avenues including, but not limited to, safety committees, incentive programs, an open-door policy, and other communications mechanisms. The strategic plan should possess one or more mechanisms through which to capture the creative idea and one or more methods of providing feedback to the employee or offering positive reinforcement.

The employee who originally brought the ideas should be asked to give feedback as to the status of their ideas. This can be provided through the minutes of the safety committee meeting, individual notices, verbal communications through safety committee representatives, or other methods. Without feedback, the employee may perceive that his or her idea is not valuable.

Though often expressed but seldom used by many companies, a true open-door policy is one mechanism through which employees can express their ideas and concerns with regards to the workplace. Under an ideal open-door policy, employees may, at any time, enter the office area of the safety and loss prevention professional and express their ideas with regard to virtually any topic. Although many companies state that they have an open-door policy, the reality is that employees are often fearful of expressing ideas to management in fear of repercussions or embarrassment. An effective open-door policy can be one of the most effective and low-cost methods of tapping into your employees' creativity. However, if not effectively managed, the open-door policy can cause a loss of productivity and created problems in and of itself.

The safety and resource control professional must actually listen to, and effectively communicate with, the individual employee. Effective listening means removing all barriers and obstacles and permitting the employee and the safety and resource control professional to focus on exactly what the employee is saying and on the employee's specific idea. This approach will allow you to tap into the creativity of your employees.

Safety and resource control professionals should seek ideas from others and create new ideas and concepts in order permit the growth of the function to continue and accelerate into the future. There are no bad ideas—only ideas that have not been tried or tested. And be tenacious. It's easy to have someone dismiss an idea or a concept—history is full of ideas that some thought were foolish, such as electricity, computers, air travel, and automobiles. Small minds reap small rewards. Think "outside the box."

control professional to tackle more technical or difficult issues and manage the function but also can pay substantial dividends through the education and active involvement of each team member to reduced monetary costs, human losses, as well as improved morale and employee relations.

In developing the appropriate mechanisms to manage the safety and resource control function, a written plan of action is one of the initial steps. A written plan of action, not unlike a battle plan in military terms, sets forth the objective of each activity, delineates the activity as smaller manageable elements, names the responsible parties for each element of the activity, and provides target dates at which time the responsible party will be held accountable for the achievement of the particular element or activity. In order to manage this planning phase, safety and resource control professionals can use a planning document that permits the safety and resource control professional to evaluate progress toward the objective on a daily basis and also to hold the appropriate responsible party accountable for the achievement of the particular element or activity. This type of planning document can be computerized or simply completed in written form.

In developing a plan of action, all levels of the management team must be involved in developing priorities and scheduling the plan of action. This team involvement assists the management team members to buy in to the overall safety and resource control efforts of the organization. It also permits input regarding potential obstacles that could be encountered and the development of a realistic targeted time schedule given the OSHA time requirements and worksite pressures. Ranking of the various mandated safety and health programs and other safety and resource control programs that are not required by OSHA or other government agencies should be given careful attention so that the appropriate programs affording employees the maximum protection and meeting the OSHA or other government agency's target dates are given first preference.

Management team members should be advised that team members will he held accountable for the successful and timely completion of their assigned tasks and duties, as set forth under the plan of action. With the required management commitment to the safety and resource control goals and objectives by top management and management team member's buy in during the development of the plan of action, management team members should be well aware of their specific duties and responsibilities within the framework of the overall safety and resource control effort of the organization. Appropriate positive or negative reinforcement can be used to achieve this purpose in addition to appropriate disciplinary action, if necessary.

Safety and resource control professionals should be cautious when developing written safety and health, environmental, or other programs to meet compliance requirements. The methods used in the development and documentation of compliance programs are a direct reflection on the safety and resource efforts of the organization. Safety and resource control professionals should always develop written compliance programs in a professional manner. The safety and resource control professionals should also be aware that when compliance officers are evaluating the written compliance programs, this could possibly set the tone for the entire inspection or investigation.

Compliance programs should also be developed in a defensive manner. Every element of the OSHA standard should be addressed in written form, and all training and education requirements should be documented. In the event of a work-related accident or incident that places the written programs on trial, a program will be placed under a microscope and every detail scrutinized from every angle. Proper preparation, evaluation, and scrutiny when developing a written compliance program can avoid substantial embarrassment, cost, and other liabilities in the future.

The safety and resource control professional should develop a strategy and method through which to effectively manage manage the various responsible parties and programs that are taking place simultaneously within the operations. The use of an audit mechanism can be an effective tool in identifying deficiencies within a compliance program, areas in which responsibilities are not being fulfilled and permit immediate correction of the deficiency. Although there are various types of safety and resource control audit instruments, all audits possess the basic elements for identification of the required elements of a individual program (i.e., tracking the current level of performance, identifying deficiencies, and identifying potential corrective actions) and of the responsible parties. A safety and resource control audit mechanism can provide numerical scoring, letter or grade scoring, or other method of scoring, so that the management team can ascertain their current level of performance and identify areas in need of improvement.

Particularly in the areas involving governmental compliance programs, it is imperative to ensure that every area of the standard or regulation has been evaluated and is in compliance. Each and every element of the specific standard or regulation must be in compliance and functioning effectively, and the individual team members' progress is assessed.

When identifying responsibilities and delegating safety and resource control job functions, it is each member of your management team must fully understand not only what their responsibilities are and how to perform the function, but also the underlying reason for the duty and responsibility. Safety and resource control professionals should provide not only the responsibility, with input from the individual and team, but also the underlying reasons and benefits, in both monetary and humanitarian terms, that can be acquired through a comprehensive and systematic management approach to the safety and resource control function. To ensure that the team fully understands the concepts involved in a proactive program, the management team should be educated in and understand how accidents happen and how accidents can be prevented. Using the domino theory, safety and resource control professionals can easily explain the causal factors leading up to a accident and the negative impact after an accident. The safety and resource control professional can explain that, through the use of a proactive safety and resource control program, the factors that could lead to an accident can be identified and corrected before the risk factors mount and ultimately lead to an accident.

The first three dominoes show the underlying factors that could lead to an accident. Safety and resource control professionals should emphasize the fact that the underlying causes for workplace injuries and illnesses can be identified and corrected through the use of a proactive safety and resource control program. If the underlying factors leading to an accident are not identified and corrected, the dominoes begin to fall; once this process is under way, it is almost impossible to prevent an accident. The key is to ensure that the management group realize that to prevent an accident, the underlying risk factors must be minimized or eliminated, rather than reacting after an accident has already happened.

To amplify this point, safety and resource control professionals often use the following progressional model to drive home the point that near misses and other underlying factors, if not addressed, will ultimately lead to an accident. In this model, for every 300 equipment

EXAMPLE

Complete audit instrument available in Appendix B

Safety and Resource Control Audit Assessment

Quarterly Report for ___________ ***Quarter of*** ___________ ***Year***

Facility Name ____

Total Points Available: XXXXX

Total Points Scored: XXXXX

Percentage Score: ____________

(Total points scored divided by total points available)

Audit Performed by: ____

Signature: ____________

Date: ________________

Management Safety Responsibilities	Answer		Total Points	Score
1. Are the safety responsibilities of each management team member in writing?	YES	NO	10	
2. Are the safety responsibilities explained completely to each team member?	YES	NO	10	
3. Does each team member receive a copy of his/her safety responsibilities?	YES	NO	5	
4. Has each team member been provided the opportunity to discuss their safety responsibilities and add input into the methods of performing these responsible acts?	YES	NO	10	
SECTION TOTAL			35	

Safety Goals	Answer		Total Points	Score
1. Has each member of the management team been able to provide input into the development of the operations safety goals?	YES	NO	5	
2. Has each member of the management team been able to provide input into their department's goals?	YES	NO	10	
3. Are goals developed in more than one safety area?	YES	NO	10	
4. Are the goals reasonable and attainable?	YES	NO	10	
5. Is there follow-up with feedback on a regular basis?	YES	NO	15	
6. Is there a method for tracking the departments progress toward their goal?	YES	NO	15	
7. Is the entire program audited on a regular basis?	YES	NO	10	
8. Does your management team fully understand purpose of the Safety Goals Program?	YES	NO	10	

9. Does your management team understand the OSHA recordable rate, loss time rate, and days lost rate (per 200,000 MANHOURS)?	YES	NO	10	
10. Does your management team fully understand the provisions and requirements when the safety goals are not achieved on a monthly basis?	YES	NO	10	
11. Is your management team provided with daily/weekly feedback regarding the attainment of their safety goals?	YES	NO	10	
Section Total			**115**	

damage accidents or near misses an employer may experience, there will be 29 minor injuries. If the deficiencies and underlying risk factors are not identified and corrected, the 300 near misses will ultimately lead to one major injury or fatality. The key is to ensure complete understanding that the management team must take a proactive approach to the safety and resource control function, rather than reacting when an incident or accident happens.

In order for most individuals or teams to embrace the concept of a proactive safety and resource control program, the individual or team, especially your management team, must be educated as to the cost-effectiveness of such an endeavor. Safety and resource control professionals are often able to show the monetary, as well as the humanitarian, benefits of a proactive safety and resource control program through the use of a cost–benefit analysis.

After your team members are educated and trained and provided the tools and have accepted their responsibilities, the safety and resource control professional must monitor each individual or team's performance and hold the individual or team accountable for their performance. Input, in the form of positive reinforcement or negative reinforcement (i.e., discipline), must be provided in a timely and effective manner. With time, continued emphasis and reinforcement, and proper management, team members will embrace their responsibilities and improve the overall safety and resource control effort.

CHAPTER 16

YOU PAY THE BILL

Money not only talks, it screams.

—Leslie B. Flynn

Fifty productive men are better than two hundred who are not.

—Talmud

Many safety and resource control professionals have found that the management and administration of their organization's workers' compensation program is part of the overall job responsibilities because the end result of work-related accidents (i.e., an injured employee) encompasses many issues regarding safety and resource control. There are many similarities between the management of a workers' compensation program and a safety and resource control program, but there are also many significant differences.

On a larger idealistic scale, workers' compensation and safety and resource control are at opposite ends of the spectrum. A workers' compensation program is generally a reactive mechanism to compensate employees with monetary benefits after an accident has occurred. Conversely, safety and resource control programs are designed by nature to be proactive programs, designed to prevent employees from being injured in the first place. Safety and resource control professionals who wear the dual hats of safety and/or health as well as

workers' compensation mus
wearing at any given time.
individuals, the situation, and
workers' compensation progran

The rising cost of workers' c
resulted in a significantly increase
Employers, always cognizant of th
workers' compensation costs have
many factors, including, but not lin
nesses, increased medical and rehabi
and benefits, and other factors. With
health professionals are often thrust ir ...orld of
workers' compensation with little or nc ...ucation regarding the rules, regulations, and requiren ... In safety and resource
control area, many of the potential liabilities encountered in the area of workers' compensation are a direct result of acts of omission rather than commission. Safety and resource control professionals should understand the basic structure and mechanics of the workers' compensation system, as well as the specific rules, regulations, and requirements under their individual state system.

Workers' compensations systems are fundamentally a no-fault mechanism through which employees who incur work-related injuries and illnesses are compensated with monetary and medical benefits. Either party's potential negligence is not an issue as long as this is the employer–employee relationship. In essence, workers' compensation is a compromise in that employees are guaranteed a percentage of their wages (generally two-thirds) and full payment for their medical costs when injured on-the-job. Employers are guaranteed a reduced monetary cost for these injuries or illnesses and are provided protection from additional or future legal action by the employee for the injury.

The typical workers' compensation system is characterized by the following features:

1. Every state in the United States has a workers' compensation system. There may be variations in the amounts of benefits, the rules, administration, and so forth, from state to state. In most states, workers' compensation is the exclusive remedy for on-the-job injuries and illnesses.

rkers' compensation is limited to employees
ed on-the-job. The specific locations as to what
s the work premises and on-the-job may vary from
o state.

egligence or fault by either party is largely inconsequential. No matter whether the employer is at fault or the employee is negligent, the injured employee generally receives workers' compensation coverage for any injury or illness incurred on the job.

4. Workers' compensation coverage is automatic; that is, employees are not required to sign up for workers' compensation coverage. By law, employers are required to obtain and carry workers' compensation insurance or be self insured.
5. Employee injuries or illnesses that "arise out of and in the course of employment" are considered compensable. These definition phrases have expanded this beyond the four corners of the workplace to include work-related injuries and illnesses incurred on the highways, at various in and out of town locations, and other such remote locals. These two concepts; "arising out of" the employment and "in the course of" the employment are the basic burdens of proof for the injured employee. Most states require both. The safety and resource control professional is strongly advised to review the case law in his/her state to see the expansive scope of these two phrases. That is, the injury or illnesses must "arise out of" that is, there must be a casual connection between the work and the injury or illness must be "in the course of" the employment; this relates to the time, place, and circumstances of the accident in relation to the employment. The key issue is a work connection between the employment and the injury/illness.
6. Most workers' compensation systems include wage-loss benefits (sometimes called time-loss benefits), usually one-half to three-fourths of the employees' average weekly wage. These benefits are normally tax free and are commonly called temporary total disability (TTD) benefits.
7. Most workers' compensation systems require payment of all medical expenses, including such expenses as hospital expenses, rehabilitation expenses, and prothesis expenses.

8. In situations in which an employee is killed, workers' compensation benefits for burial expenses and future wage-loss benefits are usually paid to the dependents.
9. When an employee incurs a injury or illness that is considered permanent in nature, most workers' compensation systems provide a dollar value for the percentage of loss to the injured employee. This is normally referred to as permanent partial disability (PPD) or permanent total disability (PTD).
10. In accepting workers' compensation benefits, the injured employee is normally required to waive any common law action to sue the employer for damages from the injury or illness.
11. If the employee is injured by a third party, the employer usually is required to provide workers' compensation coverage, but the employer can be reimbursed for these costs from any settlement that the injured employee receives through legal action or other methods.
12. Administration of the workers' compensation system in each state is normally assigned to a commission or board. The commission or board generally oversees an administrative agency located within state government, which manages the workers' compensation program within the state.
13. The Workers' Compensation Act in each state is a statutory enactment which can be amended by the state legislatures. Budgetary requirements are normally authorized and approved by the legislatures in each state.
14. The workers' compensation commission/board in each state normally develops administrative rules and regulations (e.g., rules of procedure, evidence) for the administration of workers' compensation claims in the state.
15. In most states, employers with one or more employees are normally required to possess workers' compensation coverage. Employer's are generally allowed several avenues through which to acquire this coverage. Employers can elect to acquire workers' compensation coverage from private insurance companies or from state-funded insurance programs or to become self-insured (i.e., after posting bond, the employer pays all costs directly from its coffers).

16. Most state workers' compensation provides a relatively long statute of limitations. For injury claims, most states grant a period of 1–10 years in which to file the claim for benefits. For work-related illnesses, the statute of limitations may be as high as 20–30 years from the time the employee first noticed the illness or the illness was diagnosed. An employee who incurred a work-related injury or illness is normally not required to be employed with the employer when the claim for benefits is filed.
17. Workers' compensation benefits are generally separate from the employment status of the injured employee. Injured employees may continue to maintain workers' compensation benefits even if the employment relationship is terminated, the employee is laid off, or other significant changes are made in employment status.
18. Most state workers' compensation systems possess some type of administrative hearing procedures. Most workers' compensation acts have designed a system of administrative "judges," generally known as administrative law judges (ALJ) to hear any disputes involving workers' compensation issues. Appeals from the decision of the administrative law judges are normally to the workers' compensation commission/board. Some states permit appeals to the state court system after all administrative appeals have been exhausted.

Safety and resource control professionals should be acutely aware that the workers' compensation system in every state is administrative in nature. Thus, considerable required paperwork must be completed in order for benefits to be paid in a timely manner. In most states, specific forms have been developed.

The most important form to initiate workers' compensation coverage in most states is the first report of injury/illness form. This form may be called a First Report form an application for adjustment of claim or may possess some other name or acronym, such as SF-1 or Form 100. This form, often divided into three parts so that information can be provided by the employer, employee, and attending physician, is often the catalyst that starts the workers' compensation system reaction. If this form is absent or misplaced, there is no reaction in the system, and no benefits are provided to the injured employee.

Under most workers' compensation systems, many forms need to be completed in an accurate and timely manner. Normally specific forms must be completed if an employee is to be off work or is returning to work. These include forms for the transfer from one physician to another, forms for independent medical examinations, forms for the payment of medical benefits, and forms for the payment of permanent partial or permanent total disability benefits. Safety and resource control professionals responsible for workers' compensation are advised to acquire a working knowledge of the appropriate legal forms used in their state's workers' compensation program.

In most states, information regarding the rules, regulations, and forms can be acquired directly from the state workers' compensation commission/board. Other sources for this information include your insurance carrier, self-insured administrator, or state fund administrator.

Safety and resource control professionals should be aware that workers' compensation claims have a "long tail"; that is, they stretch over a long period of time. Under the Occupational Safety and Health Administration (OSHA) recordkeeping system, familiar to most safety and resource control professionals, every year injuries and illnesses are totaled on the OSHA Form 200 log and a new year begins. This is not the case with workers' compensation. Once an employee sustains a work-related injury or illness, the employer is responsible for the management and costs until such time that the injury or illness reaches maximum medical recovery or the time limitations are exhausted. When an injury reaches maximum medical recovery, the employer may be responsible for payment of PPD partial or PTD benefits before closure of the claim. Additionally, in some states, the medical benefits can remain open indefinitely and cannot be settled or closed with the claim. In many circumstances, the workers' compensation claim for a work-related injury or illness may remain open for several years and thus require continued management and administration for the duration of the claim process.

Some states allow the employer to take the deposition of the employee claiming benefits, whereas others strictly prohibit it. Some states have a schedule of benefits and have permanent disability awards strictly on a percentage of disability from that schedule. Other states require that a medical provider outline the percentage of functional impairment caused by the injury/illness, as in the American Medical Association (AMA) Guidelines. Using this information as well as the employee's age, education, and work history, the Administrative Law Judge (ALJ) determines the amount of occupational impairment upon which permanent disability benefits are awarded. Still other states have variations on those systems.

In summary, safety and resource control professionals with responsiblities for the management of a workers' compensation program should become knowledgeable about the rules, regulations, and procedures under their individual state's workers' compensation system. Safety and resource control professionals who possess facilities or operations in several states should be aware that, although the general concepts may be the same, each state's workers' compensation program has specific rules, regulations, schedules, and procedures that may vary greatly between states. There is no substitute for knowing the rules and regulations under your state's workers' compensation system.

Safety and resource control professionals should be aware that potential liabilities in managing a workers' compensation program are many and varied. Above all, a safety and resource control professional should realize that most workers' compensation systems are no-fault systems that generally require the employer, or the employer's insurance administrator, to pay all required expenses whether the employer or employee was at fault, whether the accident was the result of employee negligence or neglect, or whether the injury or illness was the fault of another employee. Most workers' compensations systems are designed to be liberally construed in favor of the employee.

For many safety and resource control professionals who have been taught using a proactive method of identifying the underlying causes of accidents and immediately correcting the deficiency may find that management of the workers' compensation function can often be very time-consuming and frustrating and can show little progress. In situations of questionable claims, such as whether the injury of illness was actually work related, safety and resource control professionals should be aware that in many states, the employee has the right to initiate a workers' compensation claim with the workers' compensation commission/board and initiate or continue time-loss benefits and medical benefits until such time as the professional can acquire the appropriate evidence to dispute the claim benefits. This administrative procedure is often foreign to many safety and loss prevention professionals and can be stressful and frustrating to a safety and resource control professional accustomed to a more direct management style. Above all, the safety and resource control professional must realize that he or she must follow the prescribed rules, regulations, and procedures set forth under each state's workers' compensation system and any deviation thereof or failure to comply can place the company, the insurance carrier or administrator, and the safety and resource control professional at risk of potential liability.

In our modern litigious society, safety and resource control professionals should be aware that they will often be interacting with the legal profession when managing an injured employees workers' compensation claim. Although most workers' compensation systems are designed to minimize the adversarial confrontations, in many states, attorneys are actively involved in representing injured employees with their workers' compensation claims. Safety and resource control professionals should be aware that the amount of money paid by the injured employee to the attorney, generally a contingent fee, is normally set by statute within the individual Workers' Compensation Act.

Safety and resource control professionals should also be aware that when an injured employee is represented by legal counsel, often the direct lines of communication to the employee are severed and all communications must be through legal counsel. Circumvention of this communication bar by safety and resource control professionals often leads to confusion, mismanagement, and adversarial confrontations. Safety and resource control professionals should be aware of the rules and regulations of the individual state regarding contact and communication with an employee who is represented by legal counsel. One of the major components in the management of a workers' compensation program is the communications with the medical professionals who are treating the injured or ill employee. Safety and resource control professionals should be aware that this can be an area of potential miscommunication and conflict. The goal of the safety and resource control professional and the medical professional is normally the same—to make the injured employee well—but the methodology through which the goal is attained often conflicts. Although the potential liability in this area is not proscribed by statute, safety and resource control professionals should make every effort to ensure open and clear lines of communication, to avoid any such conflicts. The potential liability in this area is when there is a loss of trust between the safety and resource control professional and the medical community that can ultimately lead to additional benefit costs.

Given the many individuals who may be involved in a work-related injury situation (e.g., the injured employee, the attorney, the physician, the administrator, and the safety and resource control professional), the potential for conflict and thus litigation is relatively high. Safety and resource control professionals should know the rules and regulations of this administrative system and avoid areas of potential conflict.

The first and most common area of potential liability in the area of workers' compensation is simply not possessing or maintaining the appropriate workers' compensation coverage for employees. Often through error or omission, the employer either does not acquire the appropriate workers' compensation coverage or has allowed the coverage to lapse. In most states, the employer's failure to have the appropriate workers' compensation coverage will not deny the employee the necessary benefits. The state workers' compensation program, through a special fund or uninsured fund, will incur the costs of providing coverage to the employee but will bring a civil or criminal action against the employer for repayment and penalties. In several states, failure to provide the appropriate workers' compensation coverage can permit the individual employee to bring a legal action in addition to the legal action by the state workers' compensation agency. Often the employer is stripped of all defenses.

Given the paperwork requirements of most workers' compensation systems, safety and loss prevention professionals can incur liability for failing to file the appropriate forms in a timely manner. In most states, failure to file the appropriate forms in a timely manner can carry an interest penalty. In addition, safety and resource control professionals should be aware that it is the employee's right to file a workers' compensation claim, and it is often the employer's responsibility to file the appropriate form(s) with the agency or party. Liability can be assumed by the safety and resource control professional for refusing to file the form or failing to file the form with the agency to initiate benefits. In most states, civil and criminal penalties can be imposed for such actions, and additional penalties such as loss of self-insurance status can also be imposed on the employer.[1]

Safety and resource control professionals may be confronted with situations where it is believed that the injury or illness is not work related. Safety and resource control professionals often assume liability by playing judge and jury when the claim is being filed and inappropriately denying or delaying payment of benefits to the employee. In most states, civil and criminal penalties can be imposed for such actions and other penalties, such as loss of self-insurance status, can also be imposed on the employer. Safety and resource control professionals should become knowledgeable in the proper method through which to appropriately petition for the denial of a non-work-related claim through the proscribed adjudication process.[2]

In all states, an employee who files a workers' compensation claim has the right not to be harassed, coerced, discharged, or discriminated against for filing or pursuing the claim. Any such discrimination against an employee usually carried civil penalties from the workers' compensation agency. Often a separate civil action is permitted by the employee against the employer. In these civil actions, injunctive relief, monetary damages, and attorney fees are often awarded.

In most states, employees who file fraudulent workers' compensation claims are subject both to civil and to criminal sanctions. The employer bears the burden of proving the fraudulent claim and can often request an investigation be conducted by the workers' compensation agency. In some states, employees who intentionally fail to wear personal protective equipment or to follow safety rules can have their workers' compensation benefits reduced by a set percentage. Conversely, an employer who does not comply with the OSHA or other state safety and health regulations responsible for the injury or illness can be assessed an additional percentage of workers' compensation benefits over and above the proscribed level. With the burden of disproving that an injury or illness was incurred on the job, safety and resource control professionals are often placed in the position of an investigator or as the individual responsible for securing outside investigation services to attempt to gather the necessary information to deny a workers' compensation claim. The areas of potential liability with regard to surveillance, polygraph testing, drug testing, and other methods through which evidence can be secured is substantial.

Before embarking on any type of evidence gathering that may directly or indirectly invade the injured individual's privacy, the safety and resource control professional should seek legal counsel to identify the potential laws such as common law trespass, invasion of privacy, federal and state polygraph laws, alcohol and controlled substance testing perimeters, and other applicable laws. Potential sanctions for violations of these laws usually take the form of a civil action against the employer and individual involved, but criminal penalties can also be imposed for such actions as criminal trespass.

This is but a broad overview of the workers' compensation system for those safety and resource control professionals with direct or indirect responsibility for this important area. It is vital that safety and resource control professionals become proficient in the specific laws and regulations in their state with regard to workers' compensation and effectively manage this important reactive area.

ENDNOTES

1. *See*, for example, Kentucky Revised Statute 342.990, which proscribes

   (8) The following civil penalties shall be applicable for violations of particular provisions of this chapter; (a) Any employer subject to this chapter, who fails to make a report required by KRS 342.038, within fifteen (15) days from the date it was due, shall be fined not less than one hundred dollars ($100) not more than one thousand dollars ($1000) for each offense.

2. *See*, for example, Kentucky Revised Statute 342.990 (8)(c)(9), which states, "The commissioner shall initiate enforcement of a criminal penalty by causing a complaint to be filed with the appropriate agency."

CHAPTER 17

SAFETY AND THE LAW

> Laws are not invented; they grow out of circumstances.
>
> —Azarias

> Let every man remember that to violate the law is to trample on the blood of his father, and to tear the charter of his own and his children's liberty. Let reverence for the laws be breathed by every American mother to the lisping babe that prattles on her lap; let it be written in primers, spelling books, and almanacs. Let it be preached from the pulpit; proclaimed in the legislative halls, and enforced in the courts of justice. In short, let it become the political religion of the nation.
>
> —Abraham Lincoln

Safety and resource control professionals, and the companies they represent, face a myriad of potential civil and criminal liabilities from a wide variety of areas. In addition to the safety and environmental standards and penalties, corporate officials should also be aware of the potential liabilities with regards to other laws in the area of anti-trust and trade regulations (Sherman Act, Clayton Act, Robinson–Patman Act, Federal Trade Commission Act, and state antitrust laws), employee benefit and wage laws, federal and state tax laws, and especially the federal and state environmental laws (e.g., CERCLA, SARA, RCRA). The best protection that a company can acquire in order to avoid potential civil and criminal liability in all areas, including safety and resource control, is to ensure that its program is in compliance with the appropriate government regulations and to be able to demonstrate or prove compliance if called upon to do so.

The primary area of potential liability for safety and resource control professionals are the potential monetary fines that can be proposed by the Occupational Safety and Health Administration (OSHA) or a state plan agency in addition to the potential of other civil damages if compliance with the appropriate regulations is not acquired and maintained. Although the monetary fines proposed by OSHA and state plan agencies are often relatively small, with the sevenfold increase in the maximum fines, the possibility of six- or seven-figure fines for noncompliance is a distinct possibility. A monetary fines for noncompliance can now have a dramatic effect on the bottom line, and possibly the financial future of the company.

In our litigious society, safety and resource control professionals should realize that the potential of civil damages outside the realm of OSHA is becoming common place. Companies fact potential civil actions in a wide variety of areas including, but not limited to, products liability, discrimination in employment, and tort actions outside the exclusivity of workers' compensation. In these types of actions against a corporation or company, the total monetary expenditure, whether the case is won or lost, can be astronomical. Safety and resource control professionals should strive to educate their management team as to the "real" cost of litigation and how to avoid such litigation, through ensuring compliance with the government regulations and internal company policies and procedures.

With the increased potential of criminal sanctions being applied to a work-related injury or fatality under the OSH Act and by state prosecutors, all levels within the management hierarchy should be made aware of this possibility. Safety and resource control professionals should prepare for such catastrophes and be ready to exercise their Constitutional rights when necessary.

Acquiring and ensuring compliance is the key in avoiding much of the potential liability in the area of safety and resource control. Top-level management must be committed to creating and maintaining a safe and healthful workplace no matter what the economic or other conditions that may create difficulties. Top-level management should be actively involved in the area of safety and resource control and provide all of the necessary resources including, but not limited to, acquisition of competent personnel, providing necessary financial support, provide enforcement and moral support, in order to achieve the safety and resource control compliance objectives. In essence, management commitment and support are necessary from the top down, so that compliance can be achieved and maintained over the long run.

All levels of the management team should realize that the area of safety and resource control is constantly changing and evolving and that companies must change in order to maintain compliance with OSHA and other regulations. New standards are developed and promulgated, and the OSHRC decides cases and the courts decide issues that directly or indirectly affect companies on a daily basis. The safety and resource control professional is required to know the current status of a particular standard or law at any given time and to be in compliance. As has been said many times, ignorance of the law is no defense. Safety and resource control professionals should be aware of the changes and make the appropriate modifications to insure compliance within their operations.

Safety and resource control professionals should be aware that, in the area of compliance, proper and appropriate documentation is essential to prove that the particular program is in compliance with the applicable standard. Although many OSHA and state plan programs do not require written programs, safety and resource control professionals should be aware that the lack of supporting documentation can often affect credibility and lead to unnecessary citations. Appropriate documentation of compliance programs, required training, acquisition of equipment, and other required items removes all doubt and lends to the credibility of the company as well as the safety and resource control professional.

As with any compliance documentation, there are potential pitfalls that the safety and resource control professional should be aware of before establishing such a program. Safety and resource control audit documentation is a potential gold mine for the opposition in a civil action against your corporation because this document identified all the deficiencies within the safety and resource control program. Moreover, safety and resource control audit documentation has been requested by OSHA. If deficiencies are identified and the safety and resource control professional willfully disregards this information, once gathered, this documentation may provide "smoking gun" evidence in the event of a subsequent incident.

Safety and resource control professionals should prepare all documents with surgical preciseness and in a defensive manner. In any legal action, all documents will be viewed with hindsight because an accident or injury has already occurred and the issue of liability is now being placed with the appropriate parties. Safety and resource control professionals should prepare all documents, especially written compliance programs, company policies, and related documents, in preparation of a challenge and to ensure that the appropriate evaluations have been made of these documents by the legal department, personnel/human resource department, and/or any other applicable department before publication.

Safety and resource control professionals should be aware that the incidents of imposition of criminal sanctions, rather than monetary penalties, are usually available for use by OSHA, state prosecutors, or other government agencies, however, these criminal sanctions are generally reserved for only the most egregious and willful situations. The incidents in which OSHA has referred a matter to the Justice Department for criminal action or state prosecutors have file criminal charges are relatively small in relation to the number of incidents that happen in the United States each year. Although these cases make the headlines, if a safety and resource control professional is performing his or her job in the best way possible, the potential of criminal liability is remote at best.

Safety and resource control professionals are often a named party in civil actions against a company because of the visibility of their position, the corporate structure of the corporation, and the name recognition by employees. Bearing in mind that the ultimate goal of a civil action is to acquire monetary damages, the naming of the safety and resource control professional is often only a method through which to ensure the safety and resource control professional will be available during the discovery phase to "pick his/her brain" for information rather than actually seeking monetary damages from the individual. The ultimate "deep pocket" is the company rather than the individual, but the safety and resource control professional is often a vital link to the information and documents.

Civil liability for safety and resource control professionals in available and is often used in situations involving negligence, willful disregard, or other similar circumstances. In most situations, the company indemnifies the individual from personal liability as long as the safety and resource control professional was performing within the scope of his or her employment. If a safety and resource control professional is performing the job in a professional manner, neither civil or criminal liability is likely. That is not to say they may not be named in a lawsuit, but only that they will likely prevail on an individual basis. However, if the safety and resource control professional should breach their duty and injury or damage should result, liability may be present. Safety and resource control professionals are advised to assess their personal risks and take appropriate action to protect against the potential risk.

As discussed in Chapter 16, safety and resource control professionals should be aware that most workers' compensation laws bar civil recovery by individuals who have sustained injuries arising out of, and in the course of, their employment. With most injuries that occur as a result of a safety systems failure, workers' compensation is the sole remedy, and civil actions against the safety and resource control professional or company are usually barred. However, in situations in which willful negligence is the cause of the injury, some jurisdictions provide a separate cause of action in addition to workers' compensation. In addition, if a piece of equipment was involved in the accident, a products liability suit may be filed against the manufacturer, distributor, or supplier. There is also a good chance that the employer will be named in a third party liability suit. Any negligence on the part of the employee or his or agents may expose the company to significant liability.

Safety and resource control professionals should also be aware that the appearance or perception of fault often culminates in a legal action against the individual or corporation. Even if no civil or criminal liability is ultimately found against the safety and resource control professional or corporation, the individual or corporation can sustain immense damage in terms of legal costs, damage to reputation, and other harms to efficacy. Safety and resource control professionals should be aware of these perceptions and do everything feasible to maintain the appropriate appearance to avoid these legal action in the first place.

America has become a litigious society and the safety and resource control function is not immune from such actions. However, if a safety and resource control professional is performing his or her job in a professional manner and is doing everything possible to safeguard

EXAMPLE: Corporate Compliance Program Checklist

To assess whether your company is in need of a comprehensive compliance program in the area of safety and resource control and other areas of potential risk, the following list of questions is provided in order to assess your current position. Every "no" answer should send a signal that a potential risk is at hand for which a program is needed.

1. Does the board of directors put a high priority on safety and health, environmental, resource control and other regulatory compliance requirements?
2. Has the company adopted policies with regards to compliance with OSHA regulations and other laws having a direct bearing on the operations?
3. Has the company established and published a Code of Conduct and distributed copies to employees?
4. Has your company employed an individual(s) who will be directly responsible for safety and health and OSHA compliance? Are these individuals properly educated and prepared to manage the safety and resource control function?
5. Has the company formally developed a safety and resource control committee?
6. Does your company possess all the necessary resources to develop and maintain an effective safety and resource control compliance program?
7. Are your required safety and resource control compliance programs in writing?
8. Are corporate officers and managers and supervisors sensitive to the importance of safety and resource control compliance?
9. Are your employees involved in your safety and resource control efforts?
10. Are your corporate officers and managers committed to your safety, environmental, resource control, and other compliance efforts? Do they provide the necessary resources, staffing, and so forth, to perform the safety and resource control function successfully?
11. Does the company conduct periodic safety and resource con-

trol and other legal audits to detect compliance failures? Are deficiencies or failures corrected in a timely manner?

12. Has the company established hotlines or other mechanisms through which to facilitate reporting of safety and health, environmental, etc. compliance failures?
13. Are all employees properly trained in the required aspects of your safety and resource control compliance programs?
14. Is your training properly documented? Will your documentation prove beyond a shadow of a doubt that a particular employee was trained in a particular regulatory requirement?
15. Do you possess a new employee orientation program?
16. Does the orientation program for new employees include review of safety and resource control policies, codes of conduct, other policies and procedures?
17. Are employees provided hands-on safety and resource control training? Other no-the-job training?
18. Does your company conduct compliance training sessions to sensitive managers and rank-and-file employees to their legal responsibilities?
19. Does your company provide information and assistance to employees regarding their rights and responsibilities under the individual state's workers' compensation laws?
20. Does your company communicate safety and resource control compliance issues to employees with postings, newsletters, brochures, and manuals?
21. Does your company go beyond the bare bones compliance requirements to create a safe and healthful work environment?
22. Does your company "keep up" with the new and revised OSHA standards and other regulatory compliance requirements?
23. Does your company discipline employees for failure for follow proscribes safety rules and regulations? Is this discipline fair and consistent?
24. Is safety and resource control a high priority for your officers and directors?
25. Is your company proactive in the area of safety and resource control? Environmental? EEOC? Other compliance requirements?

the employees working for the company and comply with the various laws, the potential risks in the areas of criminal or civil liability are usually minimal. However, where the safety and resource control professional or the company is not willing or able, for whatever reasons, to provide this safe and healthful work environment, the potential risks of liability can accumulate until an incident results in the liability attaching to the individual or corporation.

CHAPTER 18

THE WIRED WORLD

> Trends are generated from the bottom up, fads from the top down.
>
> —John Naisbitt

> Whoever acquired knowledge but does not practice it is as one who ploughs but does not sow.
>
> —Sa'Di

Given the new technological changes that safety and resource control professionals have experienced in recent years, information on virtually any topic is available as close as the nearest computer. The difficulties today for safety and resource control professional typically is not a lack of information but an overload of information. The information is available on the world wide web or net. However, locating the information is often difficult without a road map.

Safety and resource control professionals should, however, become acquainted with the various "pitfalls" regarding the information which may be available at your fingertips. Is the information accurate? Is the information appropriate? And with information of a personal nature (e.g., insurance information), is the acquisition of this information vital?

Given the significant advances the world has experienced with regard to information transfer and the fact that there are very few rules or laws governing this area, safety and resource control professionals usually encounter problems at both ends of this technology spectrum. On one end of the spectrum, safety and resource control professionals need to safeguard company information and other proprietary information. A disgruntled employee or computer hacker can do significant damage to your company with a keyboard and a modern. On the other end of the spectrum, acquisition of improper information regarding team members can also create legal issues such as defamation and covert discrimination. For example, the safety and resource control professional acquires information regarding a previous back claim for a candidate who required surgery. The safety and resource control professional uses this information to eliminate the candidate from consideration for a position. Did the safety and health professional violate the Americans With Disabilities Act?

Information is power. If used properly, the new technology can be a great time saver for the safety and resource control professional as well as providing a ready storehouse of information that may not be readily available at the worksite. In order to maximize efficiency in locating safety and resource control information, the following safety, resource control, environmental, human resources, security, health, medical, compliance, equipment, and legal Web sites will assist the safety and resource control professional in the quest for information.

Eastern Kentucky University
http://www.eku.edu/fse

Kentucky Safety and Health Network, Inc.
http://www.kshn.org

Commerce Business Daily
http://www.cbd.savvy.com

Government Service Administration
http://www.pueblo.gsa.gov

Federal Money Retriever
www.idimagic.com

Federal Government Cites
http://fie.com/www/usgov.htm

World Wide Web Library
http://www.law.indiana.edu/law

Healthfinder
http://www.healthfinder.gov

ASSE:
http:flwww.ASSE.org

ASSE, Puget Sound:
http://Iwww.wolfnet.com/mroc/asse.html

ASSE ~ San Francisco:
http://www.midtown.net/sacasse

Army Industrial Hygiene:
http://chppm-ww.apges.army.mil/Armvih/

Asbestos Institute:
httu://www.odyssee.net/ai/

Ergo Web:
http:ergoweb.com

RSI/UK:
http://www.demon.co.uk/rsi

Safety Line:
http://sa ~ e.wt.com.au/safetvline/

Arbill:
http://www.arbill.com

Bradv:
http:flwww.safetyOnline.net/brady

BuilderOnline:
http://www.builderonline.com

CCOHS:
http://www.ccohs.ca/Resources/hshome.htm

Coppus:
http://www.safetyOnline.net/coppus

DuPont:
bttp://www.dupont.com

Eastman:
http://www.e∼∼stman.com

Kidde:
http://www.netpath.net/kidde

Lion Apparel:
http://www.lionapparel.com

Marshall:
http://www.marshall.com

MSA:
http://www.msanet.com

Peltor:
http://segwun.muskoka.net/erl/pelter/html

Red Cross:
http://www.redcross.org

Reg Scan:
http://www.regscan.com

Safeware:
http://www.safetyOnline.net/safeware

Seaton:
http://www.seton.com/directories.html

Strelinger:
http://www.strelinger.com

3M:
http://www.mmm.com

USC:
http://www.usc.eduldept/issm/SH.html

Uvex:
http://www.evux.com

National Safety Council:
http://www.nsc.org/nsc httn://www.cais.com/nsc

Safety Online:
http://www.safetyOnline.net

Job Stress Network:
http://www.serve.net/cse

MSDS, University of Utah:
gopher:/Aitlas.chem.utah.edu:7O/1 1/MSDS

American Health Consultants:
http://www.ahcpub.com

World Health Organization:
http://www.who.ch

Operation Safe Site:
http://www.opsafesite/com

Arthur D. Little:
http://www.adlittle.com

Cabot Safety Corp:
http://www.cabotsafetv.com

First Aid Direct:
http://www.first-aid.com

Fisher Scientific:
http://www.fisherl.com

H.L. Bouton Co. Inc:
http://www.safetyOnline.netlbouton

Lab Safety Supply:
http://www.labsafety.com

MicroClimate Systems:
http://www.microclimate.com

CMC Rescue Equipment:
http://www.cmcrescue.com/

Able Ergonomics Corp:
http://www.ableworks.com

J.J. Keller & Associates:
http://www.iikeller.com/keller.html

W.W. Grainger Inc.:
http://www.grainger.com

Pro-Am Safety Inc.:
http://www ~ pro-am.com

Vallen Safety Supply:
http://www.vallen.com

Ansell Edmont Industrial Inc.:
http://www.industry.net/ansell.edmont

Conney Safety Products Co.:
http://www.safetyOnline.net/coppus

Steel Structures Painting Council:
http://www.sspc.org

Typing Iniuries:
http://alumni.caltech.edu/dank/typin-archive.html

Compliance Control Center
http://users.aol.com/comcontrol/comply.html

Pathfinder Associates:
http://www.webcom.com/pathfinder/welcome.html

Timber Falling Consultants:
http://www.empnet.com/DentD/docs/internet.htm

Safety Directors' Home Pate:
http://www.unf.edu/iweeks/

Pennsylvania State University:
http://www.ennr.psu.edu/www/dept/arc/server/wikaerob.html

Coastal Video Communications Corp:
http://www.safetyonline.net/coastal

Occupational Safety Services Inc.:
http://www.k2nesoft.com/ossinc/

Enviro-Net MSDS Index:
http://www.enviro-net.com/msds/msds.html

University of Kansas School of Allied Health:
http://www.kumc.edu/SAH/

Seton Online Work lace Safety Information:
http://www.seton.com/safety.html

UVA's Video Display Ergonomics:
http://www.virginia.edu/enhealth/ERGONOMICS/toc.html

National Environmental Safety Compliance:
http://www.albany.net/nesc/

American Society of Mechanical Engineers:
http://www.asme.org

Canadian Centre for Occ. Health and Safety:
http://www.ccohs.ca/Resources/hshome.html

Denison University, Campus Security and Safety:
hftp://www.denison.edu/sec-safe/

Duke University Occupational and Environmental Medicine:
http://occ-env-med.mc.duke.edu/oem

National Technical Information Service:
http://www.fedworld.govntis/ntishome/html

Institute for Research in Construction:
http://www.cisti.nrc.ca/irc/irccontents.html

MSU Radiation, Chemical, & Biologcal Safety:
http://www.orcbs msu.edu

Rocky Mountain Center for Occupational and Environmental Health:
http://rocky.utah.edu

TrainingNet, Trench Safety:
http://www.auburn.edu/academic/architecture/bsc/research/trench/index.html

University of Iowa Institute for Rural & Environmental Health:
http://info.pmeh.uiowa.edu

University of London Ergonomics & Human Computer Interaction:
http://wvw.eroohci.ucl.ac.uk/

University of Virginia EPA Chemical Substance Factsheets:
http://ecosyst.drdr.virginia.edu/11/librarv/gen/toxics

Government Web Sites Search Engine:
http;//www.jefflevy.com/gov.htm

National Institutes of Health:
http://www.nih.gov

NIOSH:
http://www.cdc.gov/ncidod/EID/eid.html
http://www.cdc.gov/niosh/homepag.html

FEMA:
http://www.fema.gov/femahndex.html

OSHA:
http://www.osha.gov/

FedWorld:
http:IIwww.fedworld.gov/

CDC:
http://www.cdc.gov

NIOSH:
http://www.cdc.gov/niosh/Iiomepag.html

Federal Agencies:
http://www.lib.lsu.edu/gov/fedgov.html

FedWorld:
http://www.fedworld.com

US Department of Labor:
http://www.dol.gov/cgi-bin/consolid.pl?media+press

Federal Register:
http://www.access.gpo.gov/su docs/aces/aces/40.html

Agency for Toxic Substances and Disease Registry:
http://atsdr1.atsdr.cdc.gov:8O8O/atsdrhome.html

US Department of Health and Human Services:
http://www.os.dhhs.gov

NIST:
http://www.nist.gov/welcome.html

CHAPTER 19

KNOWING WHERE YOU ARE AND WHERE YOU WANT TO GO

Plan your work! Work your plan!

—Anonymous

In life, as in chess, forethought wins.

—Charles Buxton

It is vitally important that safety and resource control managers periodically assess themselves as well as the performance of their programs. In essence, safety and resource control managers must know where they are and their path toward the safety grail, as well as what programs and activities are functioning as expected and what programs have deteriorated or need to be instituted in order to move to the next level. Although there is no Occupational Safety and Health Administration (OSHA) requirement that requires overall evaluation and assessment of the safety and resource control function, it is important for individual managers to be able to identify the strengths and weaknesses within their programs.

Although there are many methods for which the safety and resource control manager can assess the various components and activities within their overall program, the one common demoninator is the fact that each program, policy, procedure, and activity is dissected and each element within each program is assessed and evaluated on both an individual and a group basis. Some companies use a numerical rating system, while other companies use a past/veil assessment. There is no right or wrong method with which to establish the instrumentality for this important assessment however careful consideration should be provided so that no stone is left unturned.

In developing an audit instrument through which the safety and resource control manager can properly assess the overall activities and programs, the audit should be conducted in a fair and consistent basis without any bias or politics involved. Each program, activity, policy, or other area being assessed should be analyzed to the elementary level, and the instrument questions or assessment mechanism should be clear and concise. These audit instrument should be carefully assessed and field tested before initiation within the evaluation system.

Safety and resource control managers should ensure that the audit instrument is actually performing as expected and that there are no other influences that might affect the truth and validity when the audit is being conducted within their facilities. Some influences that can affect the result of the survey instrument and the audit itself are such issues as monetary bonuses attached to the audit results, plant competition, and individual performance directly linked to the audit results. The audit results should be free of any political influences or other issues that might skew the results.

The audit instrument should include all legal requirements, compliance requirements, and other legally mandated issues as well as other safety and resource control activities encompassed within the function. The results achieved after completion of the audit should be accurate and as objective as possible. Many safety resource control managers have found over the years that, through modification, much of the subjectivity can be eliminated from the audit instrument through question modification, instrumentality structure, and other methods.

As exemplified in Chapter 15, a Safety and Resource Control Audit Assessment can be of great assistance in quantifying your program. This audit instrument uses a numerical point value on each and every element within the compliance area and other activities within the safety and resource control function. This sample audit instrument is not all-inclusive and is not the only method whereby the safety and resource control function can be appropriately evaluated and assessed. Safety and resource control professionals should strive to develop new and revolutionary methods designed to assess their programs appropriately, so that they will know their current status and keep on the path to achieve the safety grail. New technologies, software packages, and communication systems may permit an ongoing evaluation and assessment of the programs on a more timely basis than has been experienced in the past. The ultimate goal of the safety and resource control audit assessment is to know where you are at this point in time and to be able to modify your path to achieve our ultimate goals. (See Appendix B for an example of an Audit Instrument.)

CHAPTER 20

THE RAVIOLI METHODOLOGY

> A verbal contract isn't worth the paper it's written on.
>
> —Samuel Goldwin

> The value of all things contracted for is measurable by the appetite of the contractors, and there the just value is that which they be contented to give.
>
> —Thomas Hobbes

Sometimes managing the safety and resource control function can be compared with making ravioli. You look at me strangely and say: "how is this?" Have you ever made ravioli? To make ravioli you need a number of different items, a substantial period of time, and a particular expertise. When you are finished, you have an exceptional product; however, you have spent a substantial period of time and utilized a number of resources to make the ravioli. The same is with safety and resource control. You possess the tools and expertise, but often do not have the time to assemble and implement a program. So what do you do? You simply go out to the store and buy the ravioli. Its cheaper, faster, and almost as good as yours. The same is true of safety and resource control. Often you do not have the time to develop and implement a program, or you do not possess the particular expertise in the unique niche area. So what do you do? You go to the safety and resource control store and purchase the ravioli, that is, the particular program that is needed at the time. In essence, you contract with a subcontractor or consultant to develop, implement, train, and place all the other components within your facility.

Another reason that safety and resource control personnel go shopping for the ravioli, rather than making the ravioli themselves, is simply a matter of risk. If I am not a great ravioli maker, there is a substantial probability that after expending the time and cost of materials to assemble the ravioli, the ravioli may fall apart when I cook them. In the safety and resource control field, everyone cannot be an expert in every area. You may be stronger in some programs and you may not possess the appropriate level of expertise for other programs. To avoid wasting time and resources to develop and implement a program that may not achieve your ultimate goals, it is sometimes better to acquire the program from a proven vendor. At least when you arrive at dinner time, your ravioli will look good and provide the taste that you expect for your guest.

However, in purchasing your ravioli from a proven vendor, there are many things to be considered. Do I purchase from vendor A or from vendor B? Is the filling in the ravioli going to be meat, cheese, spinach, or something else? Do I purchase large ravioli or small ravioli? Will the ravioli be right for the sauce that I am using? And there is always the potential that the ravioli I purchased will also fall apart before dinner.

On the selection of a vendor to perform services within your safety and resource control programs, it is important to investigate to ensure that you are selecting the right vendor to provide the identified services. It is usual for the safety and resource control professional to check the background of the vendor, check with other companies the vendor has performed services for, ensure that the vendor has the appropriate insurances, and enter into a formal written contract or agreement with the vendor to identify this specific ravioli the vendor is to provide to your company. The goal of the safety and resource control professional is to ensure that when it is "dinnertime" the ravioli provided by the vendor is of exceptional quality and meets all specifications and that the guest throughly enjoys the ravioli.

Just as in the old axiom "too many cooks spoil the dinner," once the safety and resource control professional has made a selection as to the vendor who will make the ravioli, should the safety and resource control manager be in the kitchen monitoring the activities of the vendor? This has become an important issue for safety and resource control managers, given the recent court decisions with regards to control of the subcontractor or the vendor performing the services. Does the safety and resource control professional take a hands-off approach, or does the safety and resource control professional closely monitor the vendor to make sure the ravioli does not spoil? Should the safety and resource control professional directly control or manage the subcontractor or vendor's employees in the area of safety and resource control? Should the safety and resource control professional establish a liaison between the subcontractor or vendor and his or her office to identify OSHA violations?

For most safety and resource control professionals, two basic tactics can be used: a complete hands-off approach with regard to the compliance efforts of the subcontractors, and exercising complete control over the subcontractors.

The reason for these two distinct approaches is due to the decisions made by the OSHRC in 1976 in two companion cases. In *Anning-Johnson Company*,[1] a subcontractor who engaged in the installation of acoustical ceilings, drywall systems, and insulation was cited along with the general contractor for the total lack of perimeter guards. Although the subcontractor had complained to the general contractor about the lack of guards, no abatement had occurred before the OSHA inspection. The OSHRC found that both the subcontractor and the general contractor had responsibility for the violation. With regard to the subcontractor, the OSHRC stated, "What we are holding in effect is that even if a construction subcontractor neither created nor controlled a hazardous situation, the exposure of its employees to a condition that the employer knows or should have known to be hazardous, in light of the authority or "control" it retains over its own employees, gives rise to a duty under Section 5(a)(2) of the Act [29 U.S.C. Section 654(a)(2)]. This duty requires that the construction subcontractor do what is "realistic" under the circumstances to protect its employees from the hazard. . . ."[2]

With regard to the responsibility of the general contractor, the OSHRC held that liability for the OSHA violations would attach despite the fact that the general contractor did not have any employees at the work site. It held that the general contractor possessed sufficient control to give rise to a duty to correct the situation.[3]

In a companion case, *Grossman Steel & Aluminum Corp.*,[4] the

OSHRC reached the same result with respect to liability for OSHA violations for general and subcontractors. In this case, the OSHRC found liability for the general contractor based on the general contractor's supervisory authority and control. OSHRC stated that, "the general contractor normally has responsibility to assure that the other contractors fulfill their obligations with respect to employee safety that affects the entire site. The general contractor is well situated to obtain abatement of hazards, through either its own resources or its supervisory role, with respect to other contractors."[5]

The *Anning-Johnson/Grossman Steel* analysis derived from these companion decisions still represents the position of the OSHRC with regard to general contractors and subcontractors. Since 1976, four basic category of cases have evolved from this decision: (1) the employer's affirmative defense that the hazard was not created by nor did the employer control the worksite[6]; (2) the employer did not know of the hazard, did not possess control, and, with due diligence, could not have noticed the hazard[7]; (3) the employer either created or failed to control the hazard[8]; and (4) the employer was found to possess control over the hazard.[9] The *Anning-Johnson/Grossman Steel* analysis has been endorsed, in whole or in part, by five circuit courts of appeal.[10]

In light of *Anning-Johnson/Grossman Steel* analysis, OSHA has adopted rules for apportioning liability between the general and subcontractor on a multi-employer worksite. OSHA will hold each employer primarily responsible for the safety of his/her own employees and employers are generally held responsible for violations which their own employees are exposed, even if another contractor is contractually responsible for providing the necessary protection. However, OSHA has created a two-pronged affirmative defense through which an employer or general contractor might avoid liability. If the general contractor or employer did not create or control the hazard, liability can be avoided either by proving that the general contractor or employer took whatever steps were reasonable under the circumstances to protect the employees or that the general contractor or employer lacked the expertise to recognize the hazard.[11]

Given these decisions, the two basic approaches taken by most companies to minimize the risk of liability are asserting direct control over the worksite and ensuring compliance through the development and management of a safety and resource control program in compliance or attempts to eliminate this control or management of the worksite, thus shifting the liability solely to the subcontractor. With either approach, companies normally require the subcontractors to have in place a functioning safety and resource control program that meets the requirements and standards and to ensure compliance at the worksite. Companies should be aware, however, that relinquishing control over the work site may create other difficulties and may not serve to protect the company with regards to other areas of potential liability.

Safety and resource control professionals should also be aware of the recent decision in *Secretary of Labor* vs. *IBP, Inc.*[12] This case may have a bearing on whether of not the safety and resource control professional makes or purchases "ravioli."

LIABILITY ON MULTI-EMPLOYER WORKSITES

By Stephen C. Yohay and Arthur G. Sapper

One of the most vexing issues under the Occupational Safety and Health Act has been whether one employer on a multiemployer worksite may be liable for violations committed by another. This issue arose first in the construction industry, where OSHA sought to penalize so-called "controlling" employers, such as general contractors, for violations committed by subcontractors. The issue has also lately arisen in the industrial environment, where OSHA has sought to extend the doctrine and hold factory owners responsible for violations committed by contractors.

The principles governing liability on multiemployer worksites that have developed under the OSH Act have confused nearly everyone involved. Whether an employer will be held to be in "control," hence liable for another employer's violation, it not at all clear from the case law. For example, the Occupational Safety and Health Review Commission, which reviews OSHA citations, once held a general contractor liable for a less-than-obvious electrical violation committed by a specialized electrical contractor. It has

Source: From *Occupational Hazards*, October, 1998, p. 28.

even held an engineer in "control" when it merely drafted, at the owner's request, a contract modification addressing a safety issue.

Moreover, OSHA's enforcement policy and the commission's doctrine leave many industrial employers who hire contractors with a dilemma. OSHA tells them that, to avoid a citation, they must closely watch over the safety of their contractors' actions and even threaten to terminate contracts or expel contractor employees if violations occur. But actively overseeing a contractor has the effect of painting the host employer as exercising control, hence, liable in tort if a contractor's employee is injured. Also, close oversight of a contractor's safety conduct can actually increase one's OSHA liability. Case law developed by the Review Commission and the courts indicates that the more involved with contractor safety that a host or higher tier employer becomes, the more vulnerable to citation that employer becomes if the contractor commits a violation. In short, some case law gives an incentive to host employers to keep contractors at arm's length while other case law appears to require the very involvement that gives rise to additional liability.

Some recent OSHA standards might appear to provide concrete guidance as to the obligations of host employers *vis-à-vis* their contractors. For example, paragraph (f) of the lockout/tagout standard (29 CFR 1910.147) requires hosts and contractors to exchange certain kinds of information. Other examples can be found in paragraphs (c)(8)–(9) of the confined spaces standard (1910.146), paragraph (h) of the chemical process safety standard (1910.119) and paragraph (b)(1) of the hazardous waste standard (1910.120). But as discussed below, OSHA has lately been attempting to use the "control" doctrine to unpredictably impose a layer of additional requirements even when a host employer has complied with a standard's specific host–employer provisions.

The idea that one employer may be cited for violations committed by another is purely judge-made law. Nothing in the OSH Act provides for such liability. It has instead been the Review Commission and the federal appellate courts that have invented the principle that one employer may be cited for violations committed by another. This view is not unanimous, however. The U.S. Court of Appeals for the Fifth Circuit held in 1981 in *Melerine v. Avondale Shipyards Inc.*, 659 F.2d 706 (5th Cir. 1981) that the OSH Act does not impose such liability. And former Commission Chairman

Robert Moran long ago warned that the doctrine imposed "broad nebulous principles" rather than clear rules of conduct. For many years, however, these voices were nearly entirely ignored.

Recently, however, two decisions of the U.S. Court of Appeals for the District of Columbia Circuit have pointedly called the multi-employer doctrine into question. In *IBP Inc. v. Herman*, 144 F.3d 861 (D.C. Cir. 1998) and, before that, in *Anthony Crane Rental Inc. v. Reich*, 70 F.3d 1298 (D.C. Cir. 1995), the D.C. Circuit made clear its skepticism that the doctrine has any legal basis. Although the court decided both cases on different grounds and did not reach the validity of the multi-employer doctrine, it questioned whether the doctrine is consistent with the OSH Act and regulations adopted under it. In particular, the court questioned whether the doctrine, if it has any validity, may be applied in the non-construction environment.

We will now trace the history of the multi-employer doctrine, and discuss the recent D.C. Circuit decisions that have injected an unprecedented degree of uncertainty into its validity. We also will explain why these decisions are likely to stimulate further challenges to the legal basis of the doctrine.

Background

The OSH Act does not expressly impose liability on an employer for conditions to which his own employees are not exposed. On the contrary, it indicates that Congress intended to confine responsibility to the employment relationship. For example, the OSH Act's General Duty Clause (29 U.S.C. 654(a)(1)) states that an employer is required to protect "his" employees. The Act's variance provisions (29 U.S.C. 655(d) and (b)(6)(A)) state that an employer may obtain a variance from a standard if other safety methods provide equally safe employment to "his" employees. And the statute's repeated uses of the terms "employer" and "employee" make no sense if an employment relationship is not a condition of liability. Just as one is a "parent" only in relation to one's own children, a person is an "employer" only in relation to one's own employees.

For these reasons, the Review Commission's early case law held that an employer is not responsible for the safety of other employers' employees. Thus, in *Martin Iron Works Inc.*, 2 BNA OSHC

1063 (Rev. Comm'n 1974), the commission held that an employer who created a violative condition to which his own employees were not exposed was not liable.

This rule changed soon after the Second Circuit decided the case of *Brennan v. OSHRC* (Underhill Construction Corp.), 513 F.2d 1032 (2d Cir. 1975), which concerned a construction contractor who had, contrary to a standard, placed materials near floor edges above employees of other employers working on lower levels. The court held that, even if none of the contractor's employees were exposed, it could be cited because it was "engaged in a common undertaking" with the other construction contractors and was "in control of an area, and responsible for its maintenance." The court distinguished the duty to comply with OSHA's standards from the duty to comply with the OSH Act's General Duty Clause. Although the General Duty Clause states that an employer is required to protect "his" employees, the court observed that the clause requiring compliance with OSHA's standards (29†U.S.C. ß†654(a)(2)) states only that an employer must "comply" with standards and is not expressly limited to protection of one's own employees.

The Second Circuit cited no provision of the Act or its legislative history to support the imposition of this duty; instead, it relied on the general "remedial" nature of the OSH Act. It appeared unaware of the wording of the variance provisions of the Act. Later court cases followed the Second Circuit in imposing extra-employment liability, but they too did not identify the statutory source from which this duty came.

The Review Commission soon followed the *Underhill* decision and began to impose liability outside the employment relationship on general construction contractors, whom it held were in control of construction sites. In the *Grossman Steel* and *Anning-Johnson* cases, the commission held that general contractors have the "supervisory capacity to prevent or require the abatement of violations." Like the Second Circuit, it justified the imposition of liability by stating that construction sites have a number of contractors "whose employees mingle throughout the site," that "a hazard created by one employer can foreseeably affect the safety of employees of other employers," and that safety on a construction site can best be achieved if "each employer is responsible for assuring that its own conduct does not create hazards to any employees on the

site. . . ." The newly announced doctrine, the commission stated, is "required" by the "unique" and "peculiar" nature of the "multi-employer worksite common to the construction industry."
Later, the commission began applying an even broader multi-employer doctrine. In *Harvey Workover, Inc.*, 7 BNA OSHC 1687 (Rev. Comm'n 1979), the commission abandoned the limitation to construction sites because, it stated, employees of employers can mingle at nonconstruction sites too and be exposed to hazards created by other employers. In that case, the cited employer had created the violative condition, an oxygen-deficient atmosphere in its boat, and the employees of the two employers worked together in what might be considered a "common undertaking".
In 1995, however, the D.C. Circuit, voiced concern over the legality of this doctrine. In *Anthony Crane*, it noted that the doctrine appeared inconsistent with a regulation (1910.12) that OSHA had adopted in the very earliest days of the OSH Act. That regulation requires a construction employer to protect "each of his employees engaged in construction work" by complying with the construction standards. The court decided the case on another ground, however, and expressly reserved ruling on the issue.

Commission's Decision

Soon thereafter, the evolution of the multiemployer doctrine took an unexpected turn when the Review Commission decided *IBP Inc.*, 17 BNA OSHC 2073 (Rev. Comm'n 1979), rev'd, 144 F.3d 861 (D.C. Cir. 1998).

The case arose out of the following facts: IBP had hired DCS Sanitation Management to clean its meatpacking plant at night, when IBP's employees were not around. DCS cleaned machinery in a way that violated OSHA's lockout/tagout standard and endangered only DCS's employees. IBP's contract permitted IBP to exclude DCS from the site for OSHA violations. IBP could not directly discipline DCS employees, but it could complain to DCS management. IBP's lockout program required outside contractors to use IBP's lockout program, and stated that IBP would enforce its program against contractors, if necessary, by suspending contract services.

After a DCS employee was killed in an accident resulting from a failure to lock out a machine he was cleaning, OSHA issued citations to both DCS and IBP. OSHA alleged that IBP had violated the lockout standard in two respects: that DCS employees did not set controls of machines they were cleaning so as to turn off power to them and that IBP had "failed to enforce" lockout procedures against DCS.

The Review Commission's decision held that host employers are liable for contractor violations that they "control," even if the violations do not endanger the host's employees. It held that plant owners, like general contractors on construction sites, are required to "take reasonable steps to induce" the contractor to alter the conduct of its employees. The commission held that IBP was in "control" because, as "sole owner of the plant," it had "exclusive control over who entered and worked there," it "owned the hazardous equipment," and it "had the contractual authority to bar entry" to DCS employees violating lockout requirements. IBP was required to "exercise all of its control" to "ensure" that DCS employees were not exposed to known hazards. The commission then found that IBP did not implement all of the authority it had, particularly suspension of the contract. It also found that IBP knew of the lockout violations by DCS, could have suspended or terminated the contract and could, as an initial step, have threatened to expel DCS employees observed repeatedly violating the lockout rule.

Importance of Decision

The commission's decision stripped away all but one of the limitations and rationales that earlier commissioners had given for the doctrine. Only "control" remained. No longer did the commission focus on whether the employers were engaged in a "common undertaking" or whether a host employer was "responsible for [an area's] maintenance." The decision turned on essentially one factor: that, as a factory owner, IBP could exercise control by threatening to expel the contractor or his employees from the plant. The effect of the decision would have been to require host employers to closely supervise their contractors, even if the host employer had nothing to do with the operation and his employees were not endangered.

Until this decision, host employers could generally follow one of two approaches to deal with independent contractors. First, they could closely supervise their contractors for the sake of the contractors' employees. This approach runs the risk, however, that a host employer might be held liable in tort for incomplete supervision of the contractor. For example, in *Owens v. Process Industries*, 722 F. Supp. 70 (D.Del. 1989), a factory owner was held liable to an injured contractor employee under a "control" theory because it conducted safety inspections and required the contractor to follow its safety directions. Second, a host employer could hold independent contractors at arm's length, not supervising their work for the benefit of the contractor's employees and intervening only if their own employees or property were threatened.

The commission decision in *IBP* would have effectively removed that choice for nearly all employers, and would have increased a host's exposure to suit by injured contractor employees. Moreover, the commission's decision would have enhanced OSHA's ability to dictate relations between host employers and contractors. It would have permitted OSHA to freely second-guess whether a host's level of supervision is "reasonable."

The commission's decision would also have permitted OSHA to ignore specific provisions in a standard regulating the relations between host employers and contractors. For example, paragraph (f)(2) of the lockout/tagout standard requires the host employer to inform himself of the contractor's lockout procedures and to ensure that his employees comply with the contractor's lockout program. It does not require that the host employer oversee the contractor's compliance with lockout rules. The commission's imposition of an additional general duty of supervision would have effectively negated the limitations in those specific provisions and made an end run around the rule-making process. It would have turned the "control" doctrine into a sort of regulatory caulk that OSHA enforcement personnel could use to fill the holes in any host–contractor provision in a standard—holes that OSHA standard-writers may have deliberately created.

For example, the multi-employer provisions in the lockout standard were the subject of a regulatory impact analysis. OSHA estimated that the hours and costs associated with those provisions would range from 44,400 hours and $710,400 annually for small manufacturing firms to 75,600 hours and $1,209,600 annually for large ones, assuming it would take 15 seconds per lockout event to notify employees and 30 seconds to notify outside contractors, and that outside contractors were involved in 10 percent of all lockout events. Applying the general duty of supervision under the commission's "control" doctrine effectively nullifies such rulemaking estimates, for it permits OSHA to use the "control" theory to freely place additional burdens on the host employer.

The D.C. Circuit's Decision

IBP appealed the commission's decision to the D.C. Circuit. Its brief argued that an employer is required to comply with standards only when his own employees are affected; hence, that the multi-employer doctrine exceeds OSHA's authority. Alternatively, the company argued that the commission erred in concluding that the company's authority to cancel the DCS contract gave it "control" over the behavior of DCS employees. A coalition of seven trade associations led by the Edison Electric Institute and the National Association of Manufacturers filed an *amicus curiae* brief principally arguing that the OSHA Act does not permit liability to be imposed outside the employment relationship. In an extraordinary move, the court, acting on its own initiative, allotted the *amicus* oral argument time over and above that allotted to IBP.

The court began its decision by noting the narrow reason that gave rise to the multi-employer liability doctrine. "The doctrine had its inception in the construction industry, where numerous contractors and sub-contractors mingle throughout a single work site. Craft jurisdictional rules typically prevent specialists of one craft from performing work in another craft. . . ." The court also suggested that the commission's heavy reliance on Harvey Workover—where the commission had first applied the multiemployer doctrine to a nonconstruction site—was unjustified because the case could have been decided on a different and much narrower ground.

After stating that the doctrine "has somewhat of a checkered history," the court, citing *Anthony Crane*, made the following pointed observation: "We see tension between the Secretary's multi-employer theory and the language of the statute and regulations, and we have expressed doubt about its validity before." The court stated, however, that "it is once again unnecessary to decide that issue, because even under the expansive" doctrine applied by the commission, OSHA failed to show control by IBP.

The court found a lack of control because IBP's right to terminate the contract with the contractor did not mean that IBP could exercise control over the lockout behavior of the contractor's employees. It stated that the commission's view that IBP's ability to terminate its contract with DCS "subsumes the power to discipline individual DCS employees [would] take the meaning of 'control' to an unacceptably high level of abstraction." The most that IBP could be expected to do, the court noted, was point out safety violations to DCS supervisors and management.

Then, in a remark that will likely be important to future litigation in the construction industry, the court stated that the commission decision "seems to reduce general contractors' incentive to advance workplace safety." Instead of "cracking down on safety through contract termination," the court observed, general contractors "would respond to it simply be eliminating any reference to safety in subcontracts."

The court also rejected OSHA's argument that IBP had taken upon itself responsibility to supervise its contractor's employees. In a remark that may point the way to the resolution of the dilemma discussed above, the court characterized the contract between IBP and DCS, under which IBP could inform DCS of employee violations but DCS retained authority to discipline of its employees, as reflecting IBP's "disavowal of micromanagement" over its contractor. The court's approach thus seems to preserve the option of choosing an "arm's length" approach to dealing with independent contractors. It appears to reinforce the idea that host employers may choose to minimize their exposure to multi-employer liability, at least over violations created by the behavior of, and affecting only, contractor employees, by not engaging in micro-management of contractors. It suggests that host employers may, in these circumstances, choose to permit their independent contractors to remain independent. Host employers thus have more flexibility in fashioning contractor safety programs appropriate to the needs of their business.

Some Conclusions

The *IBP* and *Anthony Crane* decisions should encourage employers cited for violations committed by other employers to vigorously argue that there is no statutory basis for such liability. The D.C.

Circuit's skeptical observations on the doctrine should be seen as almost an invitation for employers to vigorously press the arguments that the *amicus* presented in *IBP*. Host employers should also take from the court's decision the lesson that, if they so choose from a business perspective, they can minimize liability under the OSH Act and tort law by not intruding themselves heavily into contractor activities that affect the welfare of only the contractor's employees.

Whether to use contractors or not in the areas of safety and resource control is totally dependent on the circumstances. Each safety and resource control professional must assess the circumstances and decide whether to make "ravioli." Both methods work both achieve the identified goals however which method utilized is up to the individual safety and resource control professional as to what works best, given the facts and circumstances at the time.

ENDNOTES

1. *Anning-Johnson Co.*, 4 O.S.H. Cases 1193 (1976).
2. Ibid.
3. Ibid.
4. *Grossman Steel & Aluminum Corp.*, 4 O.S.H. Cases 1185 (1976).
5. Ibid, p. 1188.
6. *See*, for example, *Data Electric Co., Inc.*, 5 O.S.H. Cases 1077 (1977); *Mayfield Construction Co.*, 5 O.S.H. Cases 1877 (1977).
7. *See*, for example, *A.A. Will Sand & Gravel Corp.*, 4 O.S.H. Cases 1442 (1976).
8. *See*, for example, *Circle Industries Corp.*, 4 O.S.H. Cases 1724 (1976).
9. *See*, for example, *Dun-Par Engineered Form Co.*, 8 O.S.H. Cases 1044 (1980), petition for review filed, No. 80-1401 (10th Cir. April 17, 1980).
10. *See*, for example, *Marshall v. Knutson Const. Co.*, 566 F. 2d 596 (8th cir. 1977); *Beatty Equip. Leasing, Inc. v. Sec. of Labor*, 577 F.2d 534 (9th cir. 1978); *Brennan v. OSHRC (Underhill Const. Corp.)* 513 F.2d 1032 (2d Cir. 1975); *Central of Ga. Railroad Co. V. OSHRC*, 576 F.2d 620 (5th Cir. 1978); *N.E. Tel. & Telegraph Co. v. Sec. of Labor*, 589 F.2d 81 (1st. Cir. 1978).
11. *OSHA Compliance Field Operations Manual* at V-F.
12. 17BNA OSHC 2073 (rev. Comm'n 1997), rev'd, 144 F.3d 861 (D.C. Cir. 1998).

CHAPTER 21

KEEPING UP WITH THE JONESES

> No man prospers so suddenly as by others errors.
>
> —Francis Bacon

> We will tackle the giants and set them back because we have a better product and better know-how of the marketplace.
>
> —Thomas Casey

Given the pace of new technology and new ideas, safety and resource control professionals must continuously grow and evolve in order to simply keep from stagnating in their job function. Unlike other areas in the working world, safety and resource control is relatively new (since the OSH Act in 1970) and continues to evolve and grow virtually daily with new standards and regulations, new ideas, new technologies and new court decisions. Safety and resource control professionals should keep abreast of the constant changes and be able to grow and modify with each new issue and idea that arises.

Safety and resource control professionals must establish a continuous learning path for themselves. Whether this learning is achieved through formal courses at your local university or community college, reading professional journals, attending continuing education conferences, self-education, or other methods, the learning must never stop. In essence, we stop learning or believes we know everything about our unique profession, the level of knowledge required will simply steam roll us into the unemployment line.

Safety and resource control professionals usually must possess expertise and a level of knowledge in a number of different areas, depending on the specific job function. To maintain this competency, safety and resource control professionals should allot a specific amount of precious time each week, month, or quarter simply to re-educate themselves with the new issues and ideas. A number of safety and resource control professionals will say that they do not have the time to attend a class, go to a conference, or even read a professional journal. In our professional area, it is imperative that the safety and resource control professional "find the time" to accomplish this important task or, over time, the expertise and knowledge base will deteriorate to a point where the effectiveness is no longer present.

Let me give you an example. If the safety and resource control professional acquires a degree in 1980 and has been working in the field for approximately 19 years, what has changed during the past 19 years? Let's start with OSHA. How many new standards have been promulgated during the past 19 years? Bloodborne pathogens? Hearing conservation? Control of hazardous energy? Chemical process safety standard? Confined space entry and rescue? How many thousands of additional standards have been promulgated? How about the environmental area? Fire area? Security area? Human resource area? If the safety and resource control professional stands still in his/her education in the profession today, it is the same as sliding backward.

In short, the safety and resource control profession demands a well-qualified professional who is not only qualified but also competent in the areas of expertise. Most professionals have spend years acquiring this level of expertise but must continuously improve and update this expertise in order to maintain an adequate level of competency. Safety and resource control professionals must be continuous learners in order to keep up with the pace of the changes we are experiencing in our profession and workplace. It is called lifetime learning and is essential in our profession.

CHAPTER 22

THE EXPANSION DRAFT

I praise loudly; I blame softly.

—Catherine II

I have found, in management, that people are the most challenging part of the equation. The people side has so many dimensions that is more of an art than a science.

—Hank Garmon

One wild card in safety and resource control program is the employment fluctuations in the workplace. In today's business world, there are corporate takeovers, downsizing, rightsizing, and changing demographics. These changes in the workplace have a dramatic effect, either directly or indirectly, on the effectiveness of your safety and resource control programs. For example, if the rumor starts that your facility is going to shut down, there is a substantial probability that a number of key personnel will jump ship and go to another company. Employees may not be concentrating on their work activities; thus, the accident rate may increase, and employees may adopt a lack a dazed attitude, items within the facility may "disappear" or become damaged, and the overall team effort may disintegrate. In essence, your program may have the legs cut out from underneath it.

The changing demographics of the American population are changing the age, sex, and national origin of many employers work population. The baby boom generation is aging, there are more women in the workplace than ever, and more languages are being spoken in the workplace than ever. These changes create new opportunities for the safety and resource professional to think outside the box in creating safety and resource programs to safeguard these individuals in the workplace.

What has traditionally been within the realm of personal and human resources, namely hiring, screening, termination, and other activities that directly or indirectly affect the employment status, also have a direct bearing on the safety and resource function. In addition to more direct involvement by the safety and resource control function, the acquisition of qualified personnel, training of qualified personnel, and retaining of qualified personnel in their specific capacities is especially important in order to maintain consistency within our safety and resource control programs. Thus safety and resource control specialist must acquire the expertise to appropriately manage this important function.

The primary issue that safety and resource professional must be absolutely confident is the employment status of their employees. There are various types of legal employment status and each category possesses specialized laws and rules. These categories include the following:

- At-will employee
- Union contract employee
- Independent contractor
- Probationary employee
- Full-time employee
- Nonunion employee with handbook
- Part-time employee

Safety and resource control professionals should be familiar with the various employment status within their facilities and the rules and regulations that apply to each category. Normally, safety and resource control professionals will have specific company policies and procedures governing each employment category, as well as specific legal considerations and tax considerations for each category. It cannot be overemphasized the fact that safety and resource control professionals with responsibilities in this area must be totally competent and knowledgeable with regards to the rules and policies governing each category of employment within their operations.

At-will employees can be terminated for good cause, bad cause, or no cause at all. However, if employees are governed by a collective bargaining agreement, employment contract, or other contracts, the at-will employment status is no longer valid. Courts have created substantial exceptions to the at-will employment doctrine that safety and resource control professionals must be aware of:

- Handbook exception
- Public policy exceptions
- Oral contract exceptions
- Statutory duty exception
- Term of years

There is a substantial difference between the classification of an employee and that of an independent contractor. This status has far-reaching consequences in terms of the law and specifically under the Fair Labor Standards Act (FLSA), Federal Insurance Contribution Act (FICA), Federal Unemployment Tax Act (FUTA), and Internal Revenue Code.

The U.S. Supreme Court has determines that there is no rule or test that can determine the status of an individual as an independent contractor under the FLSA. The Internal Revenue Service (IRS) has provided some guidance as to the method of determining the independent contractor's status; however, in virtually all circumstances, the totality of the circumstances is evaluated and significant weight is provided to the following categories:

- The nature of the relationship of the services provided to the principal's business.
- The permanency of the relationship.
- The amount of individual investment in facilities and equipment.
- The opportunities for profit and loss.
- The degree of independent business organization and operations.
- The degree of independent initiative or judgment.

Although this can be a gray area for many safety and resource control professionals, a simple test can assist in making this determination. If the answer is "yes" to any of the following, the individual is most probably an employee, rather than an independent contractor:

- Does your company set the individual's work hours?
- Does your company mandate the ways of doing the job or methods by which the task is to be implemented?
- Does your company pay the individual by the hour?
- Does your company supply the tools, materials, and work area?
- Does your company supply the telephone, secretarial services, and work space?
- Does your company establish set geographical limits on the individual?

Safety and resource control professionals, as discussed in Chapter 19, are advised to exercise extreme caution in the area of independent contractors. If the individual's status is in question, assistance should be acquired from your legal counsel. Government publications are available from the U.S. Department of Labor and IRS that can provide additional guidance at no cost.

In today's workplace, safety and resource control professionals are often asked to design and manage flexible or alternative work schedules in order to accommodate the needs of the employer as well as the employees. Safety and resource control professionals may consider the development of specific policies with regard to alternative work schedules as well as methods through which to monitor and evaluate job performance.

In many organizations, safety and resource control professionals are asked to design "flex-time" schedules through which the employee possesses the discretion to start time, quit time, and hours in which the individual is actually at their workstation. Employees often work at home or in other remote locations. This is often permitted because of the new technological modifications and the types of work being performed.

Given the changing demographics of our workforce as well as the type of work being performed, many employers have moved to a compressed work week to permit employees to complete all work assignments within a nontraditional work week. These can include, but not be limited to, employees who work 10 to 12 hour days, employees who work 24 hours on shift and 24 hours off work, and other sequencing methods.

Also because of the changing demographics of the American workforce, some employees would rather work part-time in order to permit additional hours for other jobs, taking care of children, or other needs. These employees often asked to be considered on a permanent status however would prefer to work only part-time hours. This has created a new category of employee known as the "permanent part-time" employee. Safety and resource control professionals are often asked to design specific programs to address the needs of these unique employees, including issues involving health insurance and applicability of policies.

One of the newer concepts, some employers are using is that of jobsharing. In job sharing, more than one employee perform a specific job function. Employers often permit the job-sharing employees to design their schedules, job assignments, and so forth. The important factor for the employer is the completion of the job; however, this type of job-sharing is not traditional and often requires policy modifications by the safety and resource control professional.

With our aging workforce, often employees reaching retirement age would rather reduce the number of work hours and continue to work up to the permissible age of 70. Employers do not want to lose the unique expertise skills possessed by the retiring employee. To this end, employers often permit employees to reduce the number of work hours and "phase in" their retirement. Safety and resource control professionals should evaluate retirement and pension plans carefully and, if necessary, develop specific policies and procedures governing this unique type of employment status.

With the new technologies available and the common use of home computers, many employers now permit employees to perform work while at home or at various satellite locations. Given the fact that the employer no longer can eyeball employees performing their jobs, safety and resource control professionals are often required to develop specific policies and procedures governing this type of distance work.

Although the lines as to what constitutes part-time work versus full-time work are often blurred depending on the situation and employer, traditionally a part-time employee worked 20 to 25 hours per work week on a varying schedule. Part-time employees may not be eligible for health insurance, retirement benefits, and other benefits provided to permanent or full-time employees.

The traditional permanent or full-time employee worked 40 plus hours a week on a 8-hour work schedule Monday through Friday. Permanent or full-time employees were often provided a full package of benefits including, but not limited to, health insurance, retirement benefits, and other perk programs. In today's workplace, permanent employees may not be eligible for health insurance benefits for a specified period of time, or not at all, retirement plans may be individually directed, and numerous other changes may be applicable to the individual workplace. Safety and resource control professionals should be possess a complete understanding of the specific policies and procedures governing each of these areas within their workplace.

In many companies and organizations, there is a trial period during which the employee is evaluated and can be terminated without cause, at any time for any reason. This is often called the probationary period. Safety and resource control professionals are often asked to design evaluation procedures and to establish policies and procedures governing this trial period.

Safety and resource control professionals should be aware there is a growing trend in the laws as well as court decisions governing the area of privacy rights for employees. Safety and resource control professionals should exercise extreme caution in the development of policies and procedures that may affect an employee's constitutional rights or rights guaranteed under federal, state, or local laws.

Many organizations have dress requirements for safety and image purposes, specified to employees and normally a condition of employment. For example, many organizations require a suit and tie, or organizations may require safety shoes. In the area of safety, if the company policy requires specific items of clothing, the company is normally responsible for the cost of acquiring and providing said safety clothing.

The areas of hair length and the wearing of jewelry are often a controversial area. Safety and resource control professionals often establish policies for safety reasons to minimize the exposure of longer hair to machinery and to minimize exposure of jewelry that could be caught in machinery or affect products. However, safety and resource control professionals should be extremely cautious in this area, given the potential religious significance, ethnic considerations, or other significance of the hair length or jewelry.

Although many organizations do not directly address off-duty activities in policies and procedures, safety and resource control professionals should be aware that policies may indirectly affect an individual's rights in this area, such as company policies addressing an employee's right to assess a political party or run for political office. Many company policies address the issue of felony or misdemeanor arrests by employees. For example, many companies will terminate an employee who is arrested and convicted of a felony offense. Again, the safety and resource control professionals should exercise extreme caution and insure compliance with all federal, state, and local laws.

Safety and resource control professionals must be extremely sensitive to the numerous laws, including, but not limited to, ADA, ADEA, and Title VII, which may directly or indirectly impact the issues of seniority and promotion within the organization. Careful evaluation and analysis of all seniority and promotion systems is advisable to ensure that the policy does not overtly or covertly discriminate against potential candidates.

Safety and resource control professionals are advised, given the involvement with many of the safety and resource control functions, to ensure compliance with the Americans With Disabilities Act, Rehabilitation Act, and possible state disability or handicap laws.

Can a company or organization prohibit an employee from discussing or speaking on a particular topic? Safety and resource control professionals should be extremely sensitive as to any policies or procedures that might may limit or prohibit any type of expression by employees, especially outside the workplace. The freedom of expression is a constitutionally guaranteed right that has been the subject of a number of lawsuits by employees.

Many organizations have now established cultural diversity training programs and other programs designed to sensitize the employees to the various cultural and ethnic differences. Safety and resource control professionals are advised to tailor any type of cultural diversity training programs to their work force and insure that the training itself is not discriminatory. Cultural diversity training can include, but not limited to, the following:

- Gender
- Racial stereotypes
- Intercultural conceptions
- Attitudes toward differences
- Impacts of bias

One of the newer areas of responsibility for many safety and resource control professionals is that if immigration given the new laws recently enacted in the area by Congress. Safety and resource control professionals are now required to sign immigration and naturalization service Form I-9 certifying that they have examined the required documents and that the individual is either a citizen of the United States or an alien authorized to work in the United States. Safety and resource control professionals must retain the I-9 Forms for a period of at least 3 years and may not dispose of the forms until a minimum of 1 year after the individual's termination of employment.

Safety and resource control professionals should be aware that they are now responsible for asking *all* job applicants for documents to verify that they are either a U.S. citizen or an alien authorized to work in this country. The safety and resource control professional fulfills these responsibilities if the documents are examined and each document "reasonably appears, on its face, to be genuine." The new law does not require safety and resource control professionals to verify the authenticity of these documents. Employers may not simply refuse to hire individuals who appear to be foreigners and may be subject to discrimination on the basis of national origin under the Civil Rights Act of 1964.

Given the above, how can a safety and resource control professionals motivate an employee to perform a job safely? How can safety and resource control professionals motivate employees to come to work? There are a number of employee motivational theories which can be utilized by safety and resource control professionals to motivate individuals ranging from the carrot and the stick to more novel theories. Safety and resource control professionals should be aware that there is no one theory which works in every work environment. Some of the theories that can be explored include, but not limited to, the expectancy theory, the equity theory, the "self" theories (e.g., self-esteem, self-actualization, self-efficiency), motivating through money and financial incentives, and motivating through job satisfaction and goal setting.

1. Expectancy theory
2. Equity theory
3. Self-theories: Self-esteem, self-actualization, self-efficacy
4. Motivating through compensation/financial incentives
5. Motivation and goal setting
6. Motivation and job satisfaction
7. Motivation and productivity

As stated by Daniel Considine, "To get the best out of a man go to what is best in him." Safety and resource control professionals, sometimes handcuffed by various laws and regulations, must look for new and creative ways of hiring the best employees, providing the "things" the employee needs to do the job safely, and motivating the employee to give his/her best on a daily basis. Safety and resource control professionals are encouraged to "think outside of the box" to address these important issues.

CHAPTER 23

THE WALL STREET INFLUENCE

> Do not argue with the market, for it is like the weather; Though not always kind, it is always right.
>
> —Kenneth E. Walden

> Henry Ford has several times sneered at unproductive shareholders.... Well, now. Let's see. Who made Henry Ford's own automobile company possible? The stockholders who originally advanced money to him. Who makes sit possible for you and me to be carried to and from business by trains or street cars? Stockholders.... Who made our vast telephone and telegraph service possible? Stockholders.... Were stockholders all over the country to withdraw their capital from the enterprises in which they are invested, there would be a panic ... on a scale never before known.
>
> —B.C. Forbes

As most safety and resource control professionals are aware, companies have become very bottom line oriented and the influence of Wall Street has become prevalent in the day-to-day operations. Safety and resource control professionals should be aware of this influence of the influence of Wall Street, which can have a positive or negative impact on important aspects of their program, including, but not limited to, budgets, staffing, and other areas. With employees participating in 401K plans, stock, stock option plans, and other retirement programs, individual employees have become substantially more interested and are following the value of your company stock. So who actually owns your company? In many publically held companies, the share holders are the actual owners of the vast majority of the stock in your company. Your company's board of directors actually works for the shareholders, and your chief executive officer (CEO) and other upper management officials report to the board of directors. The value of your company is based upon, in whole or in part, on the value of your stock which is traded daily on one of the stock exchanges such as New York Stock Exchange, American Stock Exchange, NASDAQ, or other stock exchanges.

In addition to the number of individuals within inside and outside your company who own stock, the technology has permitted individuals to track individual stocks on a daily basis through such television programs as MSNBC and through internet services. More individual shareholders are managing their portfolios through online trading such as E-trade and other services. Whereas, in the past, individuals and employees may have placed their money in a mutual fund and checked the progress once or twice a year, individuals and employees are now monitoring the performance of individual company stocks on a daily, or even a minute-by-minute basis. Whereas, in the past, employees and individuals may have weathered the storm, individual shareholders are now more actively involved and more willing to purchase or sell an individual stock on a moment's notice.

How does this directly and indirectly affect the safety and resource control function? First, upper management is now more focused on a quarter-by-quarter basis and looking for the return on investment (ROI) for their shareholders. With this emphasis, the safety and resource control function is now placed under a microscope in order to minimize losses that could affect the bottom line. Next, given the instantaneous communication available today, a large accident, chemical releases, and other incidents will make the news immediately which may have a detrimental effect upon the individual stock price (e.g., Exxon Valdez—Exxon Oil). Third, companies are now looking for results on a quarter-by-quarter basis, rather than a year-by-year basis. Thus, the safety and resource control manager may be required to report more often and collect additional data that could affect the bottom line.

Conversely, when a company is not doing well and shareholders are jumping ship, the dollars available within the safety and resource control budgets may be detrimentally reduced. With companies downsizing, rightsizing, and merging, external factors can have a direct impact on the expected results from your safety and resource control programs. For example, if your company announces a 20% lay-off of employees, there is a high probability that the injury and illness rates within your facility will increase do to the distraction for employees, employees not focusing on their job, and questionable claims.

In essence, the shareholders have become your boss. Shareholders normally vote with their feet. Thus, if your company incurs a major accident, misses a dividend, or doesn't meet Wall Street's expectations, there is a substantial likelihood that they will sell your stock. When a number of shareholders sell your stock, the value of your company tends to deteriorate. This can have a effect on your borrowing power and your overall operations. When operations are affected detrimentally, it will have an effect on your budget and thus your overall effectiveness in the safety and resource control function.

Safety and resource control professionals—don't forget your shareholders.

CHAPTER 24

MAXIMIZE YOUR STRENGTHS AND "CUT" YOUR LOSSES

> Look well into thyself; there is a source of strength which will always spring up if thou wilt always look there.
>
> —Marcus Aurelius Antoninus

> Better to be a strong man with a weak point, than to be a weak man without a strong point. A diamond with a flaw is more valuable than a brick without a flaw.
>
> —William J.H. Boetcher

Safety and resource control professionals must know themselves as well as their program on an intimate basis. Individual safety and resource control managers must be able to identify their strengths as well as their weaknesses in order to assemble the personnel, resources, and other components necessary for successful safety and resource control program. Safety and resource control managers must analyze themselves without reserve in order to truthfully identify their areas strengths and weaknesses. This analysis should be conducted without ego. Safety and resource managers should not lie to themselves. This analysis must be conducted in a completely objective manners, and with substantial scrutiny. Some of the issues which the safety and resource control manager may wish to consider include the following:

Am I a good communicator?
Am I a good listener?
Am I a good writer?
Am I a good trainer?
Am I competent in all areas?
What are my weak areas?
Am I a good manager?
Am I technically sound with all compliance programs?
Am I a team player?
Do I consider myself cooperative?
Am I computer-literate?
Am I one of the team?
Am I comfortable in the plant?
Am I comfortable with all members of the management team?
Am I comfortable with the level of commitment provided by management?
Is my family comfortable with the hours that I work?
Am I deficient in any educational areas?
Do my personal goals correlate with my current job function?
Do my religious beliefs conflict with my job function?

These are just a few of the various areas in which the safety and resource control manager can look in the mirror and conduct a critical analysis to identify his or her strengths and weaknesses. Although numerous other areas can be alanyzed, the primary objective of this self-analysis is for the safety and resource control manager to be able to maximize personal strengths and cover areas of potential weakness.

Not every safety and resource control manager is exceptional in all areas. Once safety and resource control managers have analyzed personal strengths and weaknesses, they can surround themselves with individuals who have strengths in the areas in which they are weakest. Safety and resource control managers should search for the opposing strengths in other individuals within their staff or management team and avoid the use of "yes persons," if they are to strengthen the overall program effort.

Selecting individuals who possess strengths in your area of weakness requires setting your ego aside. After identifying personal weaknesses, safety and resource control managers should search for the opposing strength within individual candidates when hiring, promoting, or selecting individuals for appropriate functions within the overall safety and resource control programs. The primary issue that often evolves is the fact that individuals tend to gravitate to toward individuals with similar likes, strengths, and weaknesses. This has nothing to do with race, sex, color, creed, national origin. It is simply human nature. When selecting, promoting, or otherwise assembling a new management team, individuals tend to identify their own strengths within the particular candidates. Safety and resource control managers should search for the opposites for their "opposite" so that the sum of the parts has no weak links.

Another deficiency that is often identified in the American workplace is the fact that we often promote people to their own level of their incompetency and then terminate the individual when the expected performance is not achieved. For example, individual A is the top punchpress operator in our facilities. When a supervisory position becomes available, we promote the punchpress operator to the supervisory function. Individual A may not possess the skills and ability to be a successful supervisor. Over time he is not successful in his new position. The American way is to terminate the individual from the supervisor's capacity and from the company, rather than return the individual to his level of excellence as a punch press operator. Safety and resource control managers should closely analyze the individual on their team or staff to ensure that the individuals selected for these important positions not only possess the requisite skills and ability but also fit appropriately within the team. Safety and resource control managers must ensure that the individual is provided all the tools and ability to achieve maximum performance within the safety and resource control capacity.

What is the cost of terminating an employee? Are we promoting employees to their own level of incompetency? Correlating with the promoting individuals to a level where their competency is not adequate, safety and resource control managers should not permit the inadequacies or losses to mount before making a decision. Although we try with our best efforts to place the right individuals in the appropiate job functions, sometimes we make mistakes. Safety and resource control managers often permit the individual an extended period of time to correct or rectify a identified inadequacies. How long we should permit the individual to attempt to achieve competency? Although most safety and resource control managers do not like to discipline or terminate individuals within their team or management function, under certain circumstances extended inadequacies can result in irreparable harm to the overall safety and resource control program. In these types of circumstances, it is essential that the safety and resource control manager take appropriate steps to "cut the loss" prior to the harm becoming irreparable.

Safety and resource control managers should also consider when to cut the loss within the scope and breadth of their overall safety and resource control programs. In essence, some battles cannot be won. The safety and resource control manager must be aware that sometimes the continued expending of resources and time may be a lost cause. The safety and loss control professional must know when to simply stop the effort and take the loss, not unlike a stock on Wall Street that has lost 50% of its value. For example, a safety and resource control manager who would like to establish a new safety program may attempt to sell the program to upper management, but it is rejected. The additional effort to acquire the resources to initiate the program are also unavailable. The safety and resource control manager attempts several times to have the management team accept the program however they will not due to lack of funds, lack of personnel, or other reasons. The safety and resource control manager can continue to "ride the dead horse," causing additional turmoil within the program or can simply stop and cut the losses at that point in time.

The safety and resource control manager must "know thy self" as well as "know thy program" throughly and objectively. The safety and resource control manager must play to personal strengths, cover any weaknesses through the strengths of others, and establish levels or perimeters to identify when a loss must be stopped. As discussed in Chapter 10, the safety and resource control manager is often placed in a difficult position and must make difficult decisions. Knowing thy self and knowing thy program on an intimate basis will permit the safety and resource manager to make the right call.

CHAPTER 25

THE GRAPEVINE INFLUENCE

> When you have spoken the word, it reigns over you.
> When it is unspoken you reign over it.
>
> —Arabian Proverb

> For one word a man is often deemed to be wise, and for one word he is often deemed to be foolish, we should be careful indeed what we say.
>
> —Confucius

One important area that is often overlooked by safety and resource control managers is the informal "grapevine" communication system within most organizations. Although your company may have e-mail, cellular phones, memorandum formats, and other forms of formal communications, virtually every organization has the underground grapevine communication on the work floor. Information, whether formally provided or simply gossip, is often transferred throughout the workforce usually in a fairly quick manner. This information can often be false or misleading, or it can simply be distorted throughout the word of mouth transfer. For the safety and resource control function, this informal transfer of information can often be detrimental to your program and/or to the efficiency of your program. In some circumstances, the inaccurate information provided to employees can often cause emotional distress, which can lead to accidents or claims in your workplace.

Let's try a little exercise. At your next safety meeting committee meeting in which 10 or more members are present, whisper a prepared statement to the first individual at the table and ask that he or she transfer the information verbally to the next person at the table (not permitting the others to hear the conversation). Have the second committee member verbally transfer the same information to the third, and so on. After all committee members have received the information, ask the last committee member to provide the information back to you and compare what you read to the initial person with the information that the last safety committee member understood. In most circumstances, the information has been substantially modified from the original version.

This is what happens at your facility. An employee overhears a conversation and misinterprets it, an employee becomes bored and concocts a rumor, an employee "guestimates" about an issue, or for myriad of other reasons, a rumor is started. The initial information is transferred throughout your workforce and ultimately becomes distorted by the time the information gets back to you. When the rumor involves your safety resource control program or any of its components, accidents within the facility, personnel within the function, and other important issues, the safety and resource control manager must quickly nip the rumor as or clarify the information to a larger group.

Can the internal grapevine be used in a proactive manner within the safety and resource function? Generally, the answer is yes. For example, let's say you are in the process of initiating a new safety incentive program. Basic information is provided to safety committee members or small number of employees and the information flow is initiated on the grapevine. The safety and resource control manager can provide other emphasis with signs stating that "it's coming" or permit a ground swell of enthusiasm for the program. By the time the program is initiated, the employees' curiosity or interest has peaked to the point of wanting to become actively involved in the incentive effort.

However, safety and resource control managers must exercise extreme caution when providing information that can ultimately end up on the internal grapevine. For example, an employee hears the safety and resource control manager discussing an accident in the facility. The employee returns to the work area and tells another employee, who tells another employee who tells another employee, and the information is initiated on the grapevine. What may have been a minor injury to the employee as a result of the accident may be blown way out of proportion by the time the information is received by the last person on the grapevine.

Safety and resource control managers such be cognizant that there is the formal method of communication within the facilities as well as the informal grapevine method of information transfer within most facilities. Recognition that the grapevine exist and that it can have a detrimental effect upon your program is essential. Safety and resource control managers can tap into a grapevine to find out the mood, concerns, and interest of their workforce and use this information for further study to ensure accuracy and use within the overall safety and resource control effort. Your employees will be talking daily in your workplace. Make sure the information is constructive and accurate.

CHAPTER 26

THE FUTURE OF SAFETY AND RESOURCE CONTROL

> The future is like heaven—everyone exalts it but no one wants to go there now.
>
> —James Baldwin

> Neither a wise man not a brave man lies down on the tracks of history to wait for the train of the future to run over him.
>
> —Dwight D. Eisenhower

What is the future of safety and resource control? What activities will the safety and resource control professional be performing and what challenges lie ahead? Some of the areas to ponder and to prepare for during the next few years may include the following:

- We have now transformed from Generation X to Generation Y. Are employees have expectations of more than their parents and grandparents. There is an expectation for childcare, eldercare, good working conditions, good pay, and good benefits. How are we going to maintain are current workforce and attract new and qualified employees?
- Will your employees trust your company in light of the right sizing, down sizing, mergers, acquisitions, and other structural changes within your organization? How will we maintain our

production, quality, and safety as well as managing our total resources in light of these various external changes?

- OSHA is currently addressing such issues as workplace violence and tuberculosis. What are the new areas that the safety and resource professional may encounter over the next 20 years? Industrial terrorism? Chemical sensitivities? Airborne contaminates?
- The Environmental Protection Agency (EPA) is now addressing issues on a global magnitude as well as specific industrial issues. The environment, both inside your facility and outside your facility, has become extremely important. What can we expect in the area of environmental compliance? Indoor air quality? Global issues?
- With the emergence of ISO, will there be shift in priorities of the safety and resource control professional? Will there be new duties and responsibilities with regards to ISO for the safety and resource control professional?
- With the aging of the baby boom generation, where will the safety and resource control professional be able to find an adequate workforce? What will the salary demands escalate to? Will the balance of power shift?
- Where will the technology take us in the next 20 years? Will this technology provide a better or worse environment for the safety and resource control professional?
- What new laws are on the horizon that will have a direct or indirect impact on the safety and resource control function? Will discrimination laws become stronger or weaker? Will the Employment At Will Doctrine survive?
- What security will be required to safe guard our employees and physical assets? Will our plants become isolated entities safeguarded from the outside world?
- In screening potential employees in the future, will our testing become more stringent? Will we be performing DNA testing? What drugs will be testing for?
- How will we protect the intellectual property of our company? How will we prevent individuals from walking out the door with thoughts, ideas, and other information?
- How will we train our employees? Will training be done through the Internet or some new source?

In summation, the future for safety and resource control professionals is extremely bright given the demand by companies to protect a large myriad of resources on various levels throughout the organization. However, safety and resource control professionals must realize that the world is changing and the methodology and means through which we accomplished our objectives in the past may not work in the future. Safety and resource control professionals are encouraged to think "outside the box" and anticipate major changes within the function in the very near future. The world has changed and safety and resource control professionals must change with it in order to be successful. New ideas, new methods, and new ways of doing things are essential if we are to achieve the safety and resource control grail and hold onto it over the next 20 years.

APPENDIX A

EMPLOYEE WORKPLACE RIGHTS

INTRODUCTION

The Occupational Safety and Health (OSH) Act of 1970 created the Occupational Safety and Health Administration (OSHA) within the Department of Labor and encouraged employers and employees to reduce workplace hazards and to implement safety and health programs.

In so doing, this gave employees many new rights and responsibilities, including the right to do the following.

OSHA STANDARDS AND WORKPLACE HAZARDS

Review copies of appropriate standards, rules, regulations, and requirements that the employer should have available at the workplace. Request information from the employer on safety and health hazards in the workplace, precautions that may be taken, and procedures to be followed if the employee is involved in an accident or is exposed to toxic substances. Have access to relevant employee exposure and medical records. Request that the OSHA area director conduct an inspection if they believe hazardous conditions or violations of standards exist in the workplace. Have an authorized employee representative accompany the OSHA compliance officer during the inspection tour. Respond to questions from the OSHA compliance officer, particularly if there is no authorized employee representative accompanying the compliance officer on the inspection "walkaround." Observe any monitoring or measuring of hazardous materials and see the resulting records, as specified under the Act, and as required by OSHA standards. Have an authorized representative, or themselves, review the Log and Summary of Occupational Injuries (OSHA No. 200) at a reasonable time and in a reasonable manner. Object to the abatement period set by OSHA for correcting any violation in the citation issued to the employer by writing to the OSHA area director within 15 working days from the data the employer receives the citation. Submit a written request to the National Institute for Occupational Safety and Health (NIOSH) for information on whether any substance in the workplace has potentially toxic effects in the concentration being used, and have their names withheld from the employer, if so requested. Be notified by the employer if the employer applies for a variance from an OSHA standard, and testify at a variance hearing, and appeal the final decision. Have their names withheld from employer, upon request to OSHA, if a written and signed complaint is filed. Be advised of OSHA actions regarding a complaint and request an informal review of any decision not to inspect or to issue a citation. File a Section 11(c) discrimination complaint if punished for exercising the above rights or for refusing to work when faced with an imminent danger of death or serious injury and there is insufficient time for OSHA to inspect; or file a Section 405 reprisal complaint (under the Surface Transportation Assistance Act (STAA)).

Pursuant to Section 18 of the Act, states can develop and operate their own occupational safety and health programs under state plans approved and monitored by Federal OSHA. States that assume responsibility for their own occupational safety and health program must have provisions at least as effective as those of Federal OSHA, including the protection of employee rights. There are currently 25 state plans. Twenty-one states and two territories administer plans covering both private and state and local government employment; and two states cover only the public sector. All the rights and responsibilities described in this booklet are similarly provided by state programs. (See list of these states at the end of this booklet.)

Any interested person or groups of persons, including employees, who have a complaint concerning the operation or administration of a state plan may submit a Complain About State Program Administration (CASPA) to the appropriate OSHA regional administrator (see lists at the end of this booklet.) Under CASPA procedures, the OSHA regional administrator investigates these complaints and informs the State and the complainant of these findings. Corrective action is recommended when required. OSHA Standards and Workplace Hazards

Before OSHA issues, amends, or deletes regulations, the agency publishes them in the Federal Register so that interested persons or groups may comment.

The employer has a legal obligation to inform employees of OSHA safety and health standards that apply to their workplace. Upon request, the employer must make available copies of those standards and the OSHA law itself. If more information is needed about workplace hazards than the employer can supply, it can be obtained from the nearest OSHA area office.

Under the Act, employers have a general duty to provide work and a workplace free from recognized hazards. Citations may be issued by OSHA when violations of standards are found and for violations of the general duty clause, even if no OSHA standard applies to the particular hazard.

The employer also must display in a prominent place the official OSHA poster that describes rights and responsibilities under OSHA's law.

Right to Know

Employers must establish a written, comprehensive hazard communication program that includes provisions for container labeling, material safety data sheets, and an employee training program. The program must include a list of the hazardous chemicals in each work area, the means the employer uses to inform employees of the hazards of non-routine tasks (for example, the cleaning of reactor vessels), hazards associated with chemicals in unlabeled pipes, and the way the employer will inform other employers of the hazards to which their employees may be exposed.

Access to Exposure and Medical Records

Employers must inform employees of the existence, location, and availability of their medical and exposure records when employees first begin employment and at least annually thereafter. Employers also must provide these records to employees or their designated representatives, upon request. Whenever an employer plans to stop doing business and there is no successor employer to receive and maintain these records, the employer must notify employees of their right of access to records at least 3 months before the employer ceases to do business. OSHA standards require the employer to measure exposure to harmful substances, the employee (or representative) has the right to observe the testing and to examine the records of the results. If the exposure levels are above the limit set by the standard, the employer must tell employees what will be done to reduce the exposure.

Cooperative Efforts to Reduce Hazards

OSHA encourages employers and employees to work together to reduce hazards. Employees should discuss safety and health problems with the employer, other workers, and union representatives (if there is a union). Information on OSHA requirements can be obtained from the OSHA area office. If there is a state occupational safety and health program, similar information can be obtained from the state.

OSHA State Consultation Service

If an employer, with the cooperation of employees, is unable to find acceptable corrections for hazards in the workplace, or if assistance is needed to identify hazards, employees should be sure the employer is aware of the OSHA-sponsored, state-delivered, free consultation service. This service is intended primarily for small employers in high-hazard industries. Employers can request a limited or comprehensive consultation visit by a consultant from the appropriate state consultation service. A list of telephone numbers for contacting state consultation projects is provided at the end of this booklet.

OSHA INSPECTIONS

If a hazard is not being corrected, an employee should contact the OSHA area office (or state program office) having jurisdiction. If the employee submits a written complaint and the OSHA area or state office determines that there are reasonable grounds for believing that a violation or danger exists, the office conducts an inspection.

Employee Representative

Under Section 8(e) of the Act, the workers' representative has a right to accompany an OSHA compliance officer (also referred to as a compliance safety and health officer, CSHO, or inspector) during an inspection. The representative must be chosen by the union (if there is one) or by the employees. Under no circumstances may the employer choose the workers' representative.

If employees are represented by more than one union, each union may choose a representative. Normally, the representative of each union will not accompany the inspector for the entire inspection, but will join the inspection only when it reaches the area where those union members work.

An OSHA inspector may conduct a comprehensive inspection of the entire workplace or a partial inspection limited to certain areas or aspects of the operation.

Helping the Compliance Officer Workers have a right to talk privately to the compliance officer on a confidential basis whether or not a workers' representative has been chosen.

Workers are encouraged to point out hazards, describe accidents or illnesses that resulted from those hazards, describe past worker complaints about hazards, and inform the inspector if working conditions are not normal during the inspection.

Observing Monitoring If health hazards are present in the workplace, a special OSHA health inspection may be conducted by an "industrial hygienist." This OSHA inspector may take samples to measure levels of dust, noise, fumes, or other hazardous materials.

OSHA will inform the employee representative as to whether the employer is in compliance. The inspector also will gather detailed information about the employer's efforts to control health hazards, including results of tests the employer may have conducted.

Reviewing OSHA Form 200 If the employer has more than 10 employees, the employer must maintain records of all work-related injuries and illnesses, and the employees or their representative have the right to review those records. Some industries with very low injury rates (e.g., insurance and real estate offices) are exempt from recordkeeping.

Work-related minor injuries must be recorded if they resulted in restriction of work or motion, loss of consciousness, transfer to another job, termination of employment, or medical treatment (other than first-aid). All recognized work-related illnesses and nonminor injuries also must be recorded.

AFTER AN INSPECTION

At the end of the inspection, the OSHA inspector will meet with the employer and the employee representatives in a closing conference to discuss the abatement of hazards that have been found.

If it is not practical to hold a joint conference, separate conferences will be held, and OSHA will provide written summaries, on request.

During the closing conference, the employee representative may describe, if not reported already, what hazards exist, what should be done to correct them, and how long it should take. Other facts about the history of health and safety conditions at the workplace may also be provided.

Challenging Abatement Period

Whether or not the employer accepts OSHA's actions, the employee (or representative) has the right to contest the time OSHA allows for correcting a hazard.

This contest must be filed in writing with the OSHA area director within 15 working days after the citation is issued. The contest will be decided by the Occupational Safety and Health Review Commission. The Review Commission is an independent agency and is not part of the Department of Labor.

Variances

Some employers may not be able to comply fully with a new safety and health standard in the time provided due to shortages of personnel, materials or equipment. In situations like these, employers may apply to OSHA for a temporary variance from the standard. In other cases, employers may be using methods or equipment that differ from those prescribed by OSHA, but which the employer believes are equal to or better than OSHA's requirements, and would qualify for consideration as a permanent variance. Applications for a permanent variance must basically contain the same information as those for temporary variances.

The employer must certify that workers have been informed of the variance application, that a copy has been given to the employee's representative, and that a summary of the application has been posted wherever notices are normally posted in the workplace. Employees also must be informed that they have the right to request a hearing on the application.

Employees, employers, and other interested groups are encouraged to participate in the variance process. Notices of variance application are published in the *Federal Register*, inviting all interested parties to comment on the action.

Confidentiality

OSHA will not tell the employer who requested the inspection unless the complainant indicates that he or she has no objection.

Review If No Inspection Is Made

The OSHA area director evaluates the complaint from the employee or representative and decides whether it is valid. If the area director decides not to inspect the workplace, he or she will send a certified letter to the complainant explaining the decision and the reasons for it. Complainants must be informed that they have the right to request further clarification of the decision from the area director; if still dissatisfied, they can appeal to the OSHA regional administrator for an informal review. Similarly, a decision by an area director not to issue a citation after an inspection is subject to further clarification from the area director and to an informal review by the regional administrator.

Discrimination for Using Rights

Employees have a right to seek safety and health on the job without fear of punishment. That right is spelled out in Section 11(c) of the Act.

The law states that the employer "shall not" punish or discriminate against employees for exercising such rights as complaining to the employer, union, OSHA, or any other government agency about job safety and health hazards; or for participating in OSHA inspections, conferences, hearings, or other OSHA-related activities.

Although there is nothing in the OSHA law that specifically gives an employee the right to refuse to perform an unsafe or unhealthful job assignment, OSHA's regulations, which have been upheld by the U.S. Supreme Court, provide that an employee may refuse to work when faced with an imminent danger of death or serious injury. The conditions necessary to justify a work refusal are very stringent, however, and a work refusal should be an action taken only as a last resort. If time permits, the unhealthful or unsafe condition should be reported to OSHA or other appropriate regulatory agency.

A state that is administering its own occupational safety and health enforcement program pursuant so Section 18 of the Act must have provisions as effective as those of Section 11(c) to protect employees from discharge or discrimination. OSHA, however, retains its Section 11(c) authority in all states regardless of the existence of an OSHA-approved state occupational safety and health program.

Workers who believe they have been punished for exercising safety and health rights must contact the nearest OSHA office within 30 days of the time they learn of the alleged discrimination. A representative of the employee's choosing can file the 11(c) complaint for the worker. Following a complaint, OSHA will contact the complainant and conduct an indepth interview to determine whether an investigation is necessary.

If evidence supports the conclusion that the employee has been punished for exercising safety and health rights, OSHA will ask the employer to restore that worker's job, earnings, and benefits. If the employer declines to enter into a voluntary settlement, OSHA may take the employer to court. In such cases, an attorney of the Department of Labor will conduct litigation on behalf of the employee to obtain this relief.

Section 405 of the Surface Transportation Assistance Act was enacted on January 6, 1983, and provides protection from reprisal by employers for truckers and certain other employees in the trucking industry involved in activity related to commercial motor vehicle safety and health. Secretary of Labor's Order No. 9–83 (48 Federal Register 35736, August 5, 1983) delegated to the Assistant Secretary of OSHA the authority to investigate and to issue findings and preliminary orders under Section 405.

Employees who believe they have been discriminated against for exercising their rights under Section 405 may file a complaint with OSHA within 180 days of the discrimination. OSHA will then investigate the complaint, and within 60 days after it was filed, issue findings as to whether there is a reason to believe Section 405 has been violated.

If OSHA finds that a complaint has merit, the agency also will issue an order requiring, where appropriate, abatement of the violation, reinstatement with back pay and related compensation, payment of compensatory damages, and the payment of the employee's expenses in bringing the complaint. Either the employee or employer may object to the findings. If no objection is filed within 30 days, the finding and order are final. If a timely filed objection is made, however, the objecting party is entitled to a hearing on the objection before an Administrative Law Judge of the Department of Labor.

Within 120 days of the hearing, the Secretary will issue a final order. A party aggrieved by the final order may seek judicial review in a court of appeals within 60 days of the final order. The following activities of truckers and certain employees involved in commercial motor vehicle operation are protected under Section 405:

Filing of safety or health complaints with OSHA or other regulatory agency relating to a violation of a commercial motor vehicle safety rule, regulation, standard, or order. Instituting or causing to be instituted any proceedings relating to a violation of a commercial motor vehicle safety rule, regulation, standard, or order. Testifying in any such proceedings relating to the above items.

Refusing to operate a vehicle when such operation constitutes a violation of any Federal rules, regulations, standards or orders applicable to commercial motor vehicle safety or health; or because of the employee's reasonable apprehension of serious injury to himself or the public due to the unsafe condition of the equipment. Complaining directly to management, coworkers, or others about job safety or health conditions relating to commercial motor vehicle operation.

Complaints under Section 405 are filed in the same manner as complaints under 11(c). The filing period for Section 405 is 180 days from the alleged discrimination, rather than 30 days as under Section 11(c).

In addition, Section 211 of the Asbestos Hazard Emergency Response Act provides employee protection from discrimination by school officials in retaliation for complaints about asbestos hazards in primary and secondary schools.

The protection and procedures are similar to those used under Section 11(c) of the OSH Act. Section 211 complaints must be filed within 90 days of the alleged discrimination.

Finally, Section 7 of the International Safe Container Act also provides employee protection from discrimination in retaliation for safety or health complaints about intermodal cargo containers designed to be transported interchangeably by sea and land carriers. The protection and procedures are similar to those used under Section 11(c) of the OSH Act. Section 7 complaints must be filed within 60 days of the alleged discrimination.

EMPLOYEE RESPONSIBILITIES

Although OSHA does not cite employees for violations of their responsibilities, each employee "shall comply with all occupational safety and health standards and all rules, regulations, and orders issued under the Act" that are applicable. Employee responsibilities and rights in states with their own occupational safety and health programs are generally the same as for workers in states covered by Federal OSHA. An employee should do the following:

> Read the OSHA Poster at the jobsite. Comply with all applicable OSHA standards. Follow all lawful employer safety and health rules and regulations, and wear or use prescribed protective equipment while working. Report hazardous conditions to the supervisor. Report any job-related injury or illness to the employer, and seek treatment promptly. Cooperate with the OSHA compliance officer conducting an inspection if he or she inquires about safety and health conditions in the workplace. Exercise rights under the Act in a responsible manner.

CONTACTING NIOSH

NIOSH can provide free information on the potential dangers of substances in the workplace. In some cases, NIOSH may visit a job-site to evaluate possible health hazards. The address is as follows:

National Institute for Occupational Safety and
Health Centers for Disease Control
1600 Clifton Road
Atlanta, Georgia 30333
Telephone: 404-639-3061

NIOSH will keep confidential the name of the person who asked for help if requested to do so.

OTHER SOURCES OF OSHA ASSISTANCE

Safety and Health Management Guidelines

Effective management of worker safety and health protection is a decisive factor in reducing the extent and severity of work-related injuries and illnesses and their related costs. To assist employers and employees in developing effective safety and health programs, OSHA published recommended Safety and Health Management Program Guidelines (*Federal Register* 54(18): 3908–3916, January 26, 1989). These voluntary guidelines apply to all places of employment covered by OSHA.

The guidelines identify four general elements that are critical to the development of a successful safety and health management program: Management commitment and employee involvement, worksite analysis, hazard prevention and control, and safety and health training. The guidelines recommend specific actions, under each of these general elements, to achieve an effective safety and health program. A single free copy of the guidelines can be obtained from the OSHA Publications Office, U.S. Department of Labor, OSHA/OSHA Publications, P.O. Box 37535, Washington, DC 20013-7535, by sending a self-addressed mail label with your request.

APPENDIX B

SAFETY AUDIT ASSESSMENT

Quarterly Report for ____________ ***Quarter of*** ____________ ***Year***

Facility Name ____

Total Points Available: XXXXX

Total Points Scored: XXXXX

Percentage Score: ____________

(Total points scored divided by total points available)

Audit Performed by: ____

Signature: ____________

Date: ______________

Management Safety Responsibilities	Answer		Total Points	Score
1. Are the safety responsibilities of each management team member in writing?	YES	NO	10	
2. Are the safety responsibilities explained completely to each team member?	YES	NO	10	
3. Does each team member receive a copy of his/her safety responsibilities?	YES	NO	5	
4. Has each team member been provided the opportunity to discuss their safety responsibilities and add input into the methods of performing these responsible acts?	YES	NO	10	
SECTION TOTAL			35	

Safety Goals	Answer		Total Points	Score
1. Has each member of the management team been able to provide input into the development of the operations safety goals?	YES	NO	5	
2. Has each member of the management team been able to provide input into their department's goals?	YES	NO	10	
3. Are goals developed in more than one safety area?	YES	NO	10	
4. Are the goals reasonable and attainable?	YES	NO	10	
5. Is there follow-up with feedback on a regular basis?	YES	NO	15	
6. Is there a method for tracking the departments progress toward their goal?	YES	NO	15	
7. Is the entire program audited on a regular basis?	YES	NO	10	
8. Does your management team fully understand purpose of the Safety Goals Program?	YES	NO	10	
9. Does your management team understand the OSHA recordable rate, loss time rate, and days lost rate (per 200,000 manhours)?	YES	NO	10	
10. Does your management team fully understand the provisions and requirements when the safety goals are not achieved on a monthly basis?	YES	NO	10	
11. Is your management team provided with daily/weekly feedback regarding the attainment of their safety goals?	YES	NO	10	
Section Total			115	

Accident Investigations	Answer		Total Points	Score
1. Is your medical staff thoroughly trained in the completion of the Accident Investigation Report?	YES	NO	5	
2. Are all supervisory personnel thorough trained in the completion of the Accident Investigation Report?	YES	NO	10	
3. Are all management team members completing the Accident Investigation Report accurately?	YES	NO	5	
4. Are the Accident Investigation Reports accurate, complete and readable?	YES	NO	10	
5. Are the Accident Investigation Reports being monitored for timeliness and quality?	YES	NO	10	
6. Are management team members receiving feedback on the quality of the Accident Investigation Report?	YES	NO	10	
7. Are management team members receiving feedback on safety recommendations identified on the Accident Investigation Report?	YES	NO	10	
8. Is your Accident Investigation Report system computerized?	YES	NO	15	
9. Is there follow-up on any items identified on the Accident Investigation Report to insure correction on the deficiency before there is a reoccurrence?	YES	NO	15	
10. Are Accident Investigation Reports being discussed in staff meetings, line meetings, or safety committee meetings?	YES	NO	10	
Section Total			100	

Supervisory Training	Answer		Total Points	Score
1. Have all supervisors been orientated to the safety system, policies, and procedures?	YES	NO	10	
2. Have all supervisors completed the job safety observations?	YES	NO	10	
3. Have all supervisors been educated in the accident investigation procedure?	YES	NO	10	
4. Have all supervisors been given a list of the personal protection equipment which his/her employees are required to wear?	YES	NO	10	
5. Have all supervisors been instructed on how to properly conduct a safety meeting?	YES	NO	10	
6. Have all supervisors been instructed on how to properly conduct a line meeting?	YES	NO	10	
7. Have all supervisors been educated in proper lifting techniques?	YES	NO	15	
8. Have all supervisors been orientated in hazard recognition?	YES	NO	15	

9. Are the supervisors conducting the near miss investigations?	YES	NO	20	
10. Do all supervisors stop employees performing unsafe acts?	YES	NO	10	
11. Are all supervisors First Aid trained?	YES	NO	15	
12. Are all supervisors CPR trained?	YES	NO	5	

Supervisory Training	Answer		Total Points	Score
13. Are all supervisors educated in the evacuation procedure?	YES	NO	10	
14. Do all supervisors know their responsibilities in evacuation?	YES	NO	10	
15. Are all supervisors aware of the safety goals?	YES	NO	10	
16. Have all supervisors developed department and line safety goals?	YES	NO	10	
17. Are all supervisors forklift qualified?	YES	NO	10	
18. Do all supervisors check his/her employees personal protective equipment daily?	YES	NO	15	
19. Do all supervisors, superintendents, and/or other management team members talk with employees regarding cumulative trauma illnesses?	YES	NO	10	
20. Are all employees educated and trained in the respiratory protection program?	YES	NO	15	
21. Have all supervisors been educated in and are completely familiar with the safety policies?	YES	NO	10	
22. Have all supervisors completed the Hazard Communication program?	YES	NO	10	
23. Are all supervisors aware of their responsibilities under the non-routine training section of the Hazard Communication program?	YES	NO	10	
Section Total			260	

Hourly Employee Training	Answer		Total Points	Score
1. Do you have a written safety orientation for new employees?	YES	NO	5	
2. Do you use audiovisual aids to help employees understand safety precautions?	YES	NO	5	
3. Do you discuss the reporting of all injuries and hazards with all employees?	YES	NO	10	
4. Have all new employees read, understand, and signed the documentation sheet for all safety policies?	YES	NO	5	
5. Does the trainer or supervisor discuss the proper use and method of wearing the required personal protective equipment?	YES	NO	10	
6. Are all safety rules and regulations discussed with all employees?	YES	NO	10	
7. Does the trainer/supervisor discuss muscle soreness and cumulative trauma illnesses with new employees?	YES	NO	10	
8. Does the trainer/supervisor recommend exercises or other techniques to assist the employee through the "breaking-in" period?	YES	NO	10	
9. Are specific job skill techniques taught?	YES	NO	15	
10. Are proper cleaning procedures taught to all new employees?	YES	NO	10	
11. Are the proper safety procedures taught to all new employees?	YES	NO	10	
12. Is the new employee receiving follow-up instruction on specific skills techniques?	YES	NO	15	

Hourly Employee Training	Answer		Total Points	Score
13. Does the supervisor/trainer discuss proper lifting techniques with each employee?	YES	NO	10	
14. Is the proper method of performing the job thoroughly explained to the new employee?	YES	NO	10	
15. Is the new employee receiving daily positive feedback from the supervisor?	YES	NO	15	
16. Is the new employee encouraged to report all "pain" to the supervisor?	YES	NO	5	
Section Total			155	

Fire Control	Answer		Total Points	Score
1. Are weekly documented inspections being conducted on the fire extinguisher?	YES	NO	10	
2. Are weekly/monthly documented inspections being conducted on all phases of the fire system?	YES	NO	10	
3. Are all fire inspection records being kept updated?	YES	NO	10	
4. Do you have a written fire plan?	YES	NO	15	
5. Do you have a notification list of telephone numbers to call in case of a fire?	YES	NO	10	
6. Do you have a fire investigation procedure?	YES	NO	5	
7. Does the maintenance department utilize the call-in procedure whenever the fire system is shut down?	YES	NO	10	
8. Do you have a designated individual thoroughly trained in the use of the fire system to conduct tours with the fire inspector, loss control personnel, etc.?	YES	NO	5	
9. Is the Safety Department being notified on all fires?	YES	NO	10	
10. Are you maintaining the required inspection documentation properly?	YES	NO	10	
Section Total			95	

Disaster Preparedness	Answer		Total Points	Score
1. Do you have a written disaster preparedness plan for your facility?	YES	NO	20	
2. Do you have a written disaster preparedness responsibility list?	YES	NO	10	
3. Do you have a written evacuation plan?	YES	NO	15	
4. Are the evacuation routes posted?	YES	NO	10	
5. Do you have emergency lighting?	YES	NO	10	
6. Is the emergency lighting inspected on a weekly basis?	YES	NO	10	
7. Do you have a written natural disaster plan? (AKA tornado, hurricane)	YES	NO	10	
8. Is there a notification list for local fire department, ambulance, police, and hospital?	YES	NO	10	
9. Are all supervisory personnel aware of their responsibilities in an evacuation?	YES	NO	10	
10. Have triage and identification areas been designated?	YES	NO	5	
11. Has an employee identification/notification procedure been developed for the evacuation procedure?	YES	NO	5	
12. Have command posts been designated?	YES	NO	5	
13. Do you have quarterly meetings to review with your disaster preparedness staff?	YES	NO	10	
14. Do you have mock evacuation drills?	YES	NO	10	
15. Do you possess a bomb threat procedure?	YES	NO	10	
16. Is your management team in full understanding of the bomb threat procedure?	YES	NO	10	
SECTION TOTAL			160	

Medical	Answer		Total Points	Score
1. Is your medical staff fully qualified?	YES	NO	10	
2. Do you have all necessary equipment on hand?	YES	NO	10	
3. Are you inventorying and purchasing necessary medical supplies in bulk in order to achieve the best price?	YES	NO	5	
4. Are you equipped for a trauma situation? (i.e., air splints, oxygen, etc.)	YES	NO	15	
5. Is your medical staff fully trained in the plant system?	YES	NO	10	
6. Is your medical staff fully trained in the individual state requirements?	YES	NO	10	

7. Does your dispensary have an emergency notification with the telephone numbers of the ambulance, hospital, etc?	YES	NO	10	
8. Is your staff fully trained in the post offer screening and physical examination procedures?	YES	NO	10	
9. Does you medical staff have daily communication with the insurance and workers compensation administrators?	YES	NO	10	
10. Does your medical staff contact local physicians and hospitals on time loss claims?	YES	NO	10	
11. Does you medical staff track all injuries and lost time cases?	YES	NO	15	
12. Does your medical staff conduct the home and hospital visitation program?	YES	NO	5	
13. Are your trauma kits inspected, cleaned, and restocked on a weekly basis?	YES	NO	10	
14. Is your medical staff involved in local medical community activities?	YES	NO	5	
15. Does your medical staff tour the plant and know each area of the plant?	YES	NO	5	
16. Is your medical staff conducting yearly evaluations on employees using the respiratory equipment?	YES	NO	10	
17. Are the first aid boxes and stretchers inspected on a weekly basis?	YES	NO	10	
18. Is your medical staff fully trained in the proper use of the alcohol/controlled substance testing equipment?	YES	NO	20	
19. Is your medical staff conducting the alcohol/controlled substance testing properly?	YES	NO	20	
Section Total			**195**	

Personal Protective Equipment	Answer		Total Points	Score
1. Do you have a list of the required personal protective equipment for each job posted? Has each job been analyzed and certified to insure the proper PPE is being worn ? Is the PPE program in compliance with the April 5, 1994 Final Ruling on PPE?	YES	NO	10	
2. Are all supervisors checking his/her employees personal protective equipment on a daily basis?	YES	NO	15	
3. Have all employees read and signed the Personal Protective Equipment Policy? Have all employees been trained as to how to properly wear the required PPE ? Have all employees signed a document stating that he/she has completed the required training and fully understands the policies and rules ?	YES	NO	10	
4. Are all supervisors insuring any employee switching jobs is wearing the proper equipment before he/she is allowed to start the job?	YES	NO	15	
5. Is the Personal Protective Equipment Policy posted?	YES	NO	10	
6. Is worn or broken equipment replaced immediately? Is there a policy within your written program specifying PPE replacement procedures ?	YES	NO	15	
7. Do all supervisory personnel council/discipline employees for not wearing the proper personal protective equipment in accordance with the policy?	YES	NO	15	
8. Do you have the necessary respiratory equipment in the plant?	YES	NO	15	
9. Do you have a written respiratory protection program? Do you possess a written PPE program ?	YES	NO	20	
10. Is the respiratory protection equipment inspected on a weekly basis? (Documented inspection)	YES	NO	10	
11. Is there annual training on the use and care of the respiratory equipment? Is annual or as-needed follow-up training being provided for all PPE ?	YES	NO	20	
12. Are all supervisors properly completing the daily inspection for PPE?	YES	NO	10	
Section Total				

Safety Committee	Answer		Total Points	Score
1. Do you have a safety committee?	YES	NO	10	
2. Does the safety committee meet on a monthly basis?	YES	NO	10	
3. Are minutes taken at each safety committee meeting?	YES	NO	10	

4.	Do you offer any educational, promotional, or informational information at these meetings? (i.e., safety meetings, statistics, literature, etc.)	YES	NO	5	
5.	Are safety committee members given a chance to discuss safety in the line meetings?	YES	NO	5	
6.	Are the items cited by the safety committee corrected and/or given an explanation why not corrected in a timely manner?	YES	NO	10	
7.	Do you have other safety related or ergonomic committees?	YES	NO	10	
8.	Do these committees meet on a periodic basis? Are written minutes of these meeting documented and maintained on a permanent basis ?	YES	NO	10	
9.	Does your management team meet on a weekly basis to review your safety goals, accidents, etc.?	YES	NO	10	
	Section Total			**80**	

	Safety Promotion	Answer		Total Points	Score
1.	Is safety being promoted on the bulletin boards?	YES	NO	5	
2.	Are safety videos being played for the hourly personnel at lunch time?	YES	NO	5	
3.	Do you use other safety promotional ideas?	YES	NO	5	
4.	Are you utilizing an incentive program of any type?	YES	NO	5	
5.	Are you using safety videotapes in your training program?	YES	NO	10	
	Section Total			**195**	

	Job Safety Analysis	Answer		Total Points	Score
1.	Have you identified your high injury areas?	YES	NO	5	
2.	Have you identified your high sprain/strain areas?	YES	NO	5	
3.	Have you identified your back injury areas?	YES	NO	5	
4.	Have you identified your potential occupational illness areas?	YES	NO	5	
5.	Have you analyzed each job for required safety equipment?	YES	NO	10	
6.	Have you analyzed each for proper safety techniques?	YES	NO	10	
7.	Have you written a job hazard/safety analysis for each job?	YES	NO	10	
8.	Have you identified the day of the week the alleged injuries most frequently occur?	YES	NO	5	
	Section Total			**55**	

	Lost Time/Restricted Duty Tracking	Answer		Total Points	Score
1.	Do you have a system to track all lost time injuries?	YES	NO	10	
2.	Do you follow-up and know the status of all lost time cases daily?	YES	NO	15	
3.	Do you track and follow-up on all employees returning to restricted duty?	YES	NO	15	
4.	Do you communicate with the attending physician and insurance administration on a daily basis on the Time Loss Claims?	YES	NO	15	
5.	Are you insuring the restricted duty individuals are performing jobs meeting the limitations of the attending physician?	YES	NO	10	
6.	Do you have a restricted duty log?	YES	NO	5	
7.	Are you following up with the attending physician on restricted duty returns?	YES	NO	10	
8.	Have you computerized your accident investigation system to identify trends?	YES	NO	10	
9.	Is the data from the computerized system being provided to your management team?	YES	NO	10	
10.	Are you analyzing the trends on your computerized system on at least a monthly basis?	YES	NO	10	
	SECTION TOTAL			**110**	

Hearing Conservation	Answer		Total Points	Score
1. Are your nurses, employment personnel, safety manager or other appropriate personnel audiometric certified?	YES	NO	10	
2. Have you analyzed your plant and taken the appropriate measures to reduce noise levels?	YES	NO	10	
3. Is your audiometric calibrated? (Note: Annual test)	YES	NO	5	
4. Have you conducted annual documented noise level surveys?	YES	NO	5	
5. Are you conducting baseline hearing tests on all new employees?	YES	NO	10	
6. Are you conducting follow-up audiometric testing?	YES	NO	5	
7. Are employees required in required areas being issued hearing protection?	YES	NO	10	
8. Are employees utilizing this hearing protection?	YES	NO	10	
9. Are supervisory personnel enforcing the hearing protection policy?	YES	NO	15	
10. Are you conducting annual audiometric testing on all employees?	YES	NO	10	
11. Are all employees trained in the proper method of wearing and caring for hearing protection?	YES	NO	10	
12. Is your program in compliance with OSHA guidelines?	YES	NO	10	
13. Is your medical staff thoroughly trained in the Impact system?	YES	NO	10	
14. Is your medical staff maintaining the hearing conservation records properly?	YES	NO	10	
15. Is your medical staff providing the Impact reports to all personnel tested?	YES	NO	10	
16. Is your medical staff working with employees and contractors to answer questions and insure compliance?	YES	NO	10	
17. Do you have a written Hearing Conservation program?	YES	NO	10	
18. Have all appropriate personnel completed the required training?	YES	NO	10	
19. Are all management team members in full understanding of the personal protective equipment policy and the appropriate disciplinary action for failure to wear hearing protection?	YES	NO	10	
20. Do you possess a copy of the OSHA Occupational Noise Exposure and Hearing Conservation Amendment?	YES	NO	10	
21. Is a copy of the OSHA Occupational Noise Exposure and Hearing Conservation Amendment posted in your plant?	YES	NO	10	
22. Is your plant management team reviewing the records and reports on a quarterly basis?	YES	NO	10	
23. Is retesting and retraining being conducted for the personnel identified by the reports?	YES	NO	10	
24. Does your management team have a good working relationship with outside contractors performing the testing (if applicable)?	YES	NO	10	
25. Do you possess a hearing conservation written program?	YES	NO	10	
26. Is your recordkeeping system and overall program in compliance?	YES	NO	10	
Section Total			**200**	

OSHA	Answer		Total Points	Score
1. Do you possess the Company's procedure guide for OSHA?	YES	NO	5	
2. Is your OSHA 200 being kept properly?	YES	NO	10	
3. Are citations posted when required?	YES	NO	10	
4. Are all required postings on bulletin boards? (i.e., General OSHA, noise, reporting or injuries, etc.)	YES	NO	10	
5. Are cameras, tape players, etc., available if needed during an OSHA visit?	YES	NO	5	
6. Has one individual been designated to act as a spokesman with OSHA?	YES	NO	5	
7. Is the spokesman aware of the proper notification procedure when OSHA arrives?	YES	NO	10	
8. Do you conduct a weekly safety inspection of your facility?	YES	NO	10	

9. Does your management team conduct a periodic planned inspection of your facility?	YES	NO	10	
10. Is your management team knowledgeable in the recordkeeping requirements for occupational injuries and illnesses?	YES	NO	25	
Section Total			**100**	

General Safety	Answer		Total Points	Score
1. Is the "Lock Out and Tagout" procedure being utilized?	YES	NO	20	
2. Have all appropriate personnel been trained in and signed the document regarding the lockout/tagout policy?	YES	NO	10	
3. Are first aid stations and stretchers inspected weekly?	YES	NO	10	
4. Is the "Welding Tag" procedure being utilized?	YES	NO	5	
5. Is your piping color-coded?	YES	NO	5	
6. Are safety glassed required in your shops?	YES	NO	10	
7. Are company electrical standards being followed?	YES	NO	10	
8. Are all employee's hand tools inspected on a regular basis?	YES	NO	5	
9. Are all company standards for ladder and scaffolding being followed?	YES	NO	10	
10. Are all company standards for stairways, floors, and wall coverings being followed?	YES	NO	10	
11. Are safety items being repaired/replaced in a timely manner?	YES	NO	15	
12. Do you have a confined entry program in writing? Does this program comply with the OSHA standard ? Are all elements of the entry and rescue procedures established ?	YES	NO	20	
13. Have all appropriate personnel been trained in and signed the document regarding the confined space entry procedure? Do you possess all required PPE and monitoring equipment ? Is this equipment in working condition and calibrated (if necessary) ?	YES	NO	10	
14. Is your confined space entry procedure being utilized?	YES	NO	20	
15. Is your confined space entry procedure training documented?	YES	NO	15	
16. Are the bottle gas procedures being followed?	YES	NO	10	
17. Are all subcontractors following the Company safety policies and standards? Does your written contracts with subcontractors address safety and health requirements ? Do you have a procedure for notifying contractors for if violations of safety policies are identified ? If this notification in writing ?	YES	NO	15	
18. Do you possess a copy of all appropriate training aids & videotapes on location?	YES	NO	10	
Section Total			**210**	

Material Hazard Identification	Answer		Total Points	Score
1. Does your plant have a written Material Hazard Communication Program?	YES	NO	15	
2. Does your plant have safety data sheets for all chemicals?	YES	NO	15	
3. Is there a list of chemicals used in the plant?	YES	NO	10	
4. Does the medical personnel have the information regarding treatment for these chemicals?	YES	NO	15	
5. Are the chemicals properly stored?	YES	NO	15	
6. Are the chemicals properly ventilated?	YES	NO	10	
7. Is fire protection available? (where needed?)	YES	NO	10	
8. Do you have emergency eyewash properly located?	YES	NO	10	
9. Do you have a chemical spill procedure? Do you have hazmat procedures in place ? Is you SARA regulated chemicals (i.e. ammonia, etc.) registered and in compliance with EPA standards ?	YES	NO	10	
10. Are you conducting periodic inspections of your facility to insure your chemical list is up to date? Do you require MSDS for all new chemicals entering the plant ?	YES	NO	10	
11. Are all chemicals, barrels, tanks, lines, etc., properly marked and labeled?	YES	NO	10	
Section Total			**130**	

Legal		Answer		Total Points	Score
1.	Is your staff knowledgeable in the areas of OSHA?	YES	NO	5	
2.	Does the medical/safety staff know the proper procedure for denying a workman's compensation claim?	YES	NO	10	
3.	Is the safety department knowledgeable in the worker's compensation laws of your state?	YES	NO	10	
4.	Do you utilize "outside" legal guidance personnel for worker's compensation claims?	YES	NO	5	
5.	Does the plant personnel have a good working relationship with outside attorneys and WC administrators?	YES	NO	5	
6.	Do you receive periodic updates on claims being handled by outside legal personnel?	YES	NO	5	
7.	Does the safety manager or designee attend all workman's compensation hearings?	YES	NO	10	
8.	Are all denial of workman's compensation benefits being approved after legal review ?	YES	NO	10	
	SECTION TOTAL			**60**	

Reference Materials		Answer		Total Points	Score
1.	Do you have a Company Safety Manual ?	YES	NO	5	
2.	Do you have an OSHA General Standards Manual on location?	YES	NO	5	
3.	Do you have a copy of the individual state's Worker's Compensation laws?	YES	NO	5	
4.	Do you have a copy of your state's safety/health codes?	YES	NO	5	
5.	Do you have a copy of the individual state's "fee schedule" or Workman's Compensation Rates?	YES	NO	5	
6.	Do you receive the monthly safety report from the Company's safety department?	YES	NO	5	
7.	Do you possess the other safety books, texts, and materials?	YES	NO	5	
8.	Do you possess the training manual ?	YES	NO	5	
9.	Do you possess the all required written compliance programs? Are your complaince programs updated on an annual basis ? Are all new standards provided a written program ? Are all modifications to current standards addressed in your written compliance programs ?	YES	NO	5	
	Section Total			**45**	

Machine Guarding		Answer		Total Points	Score
1.	Are all V-belt drivers guarded?	YES	NO	10	
2.	Are all pinch points guarded?	YES	NO	10	
3.	Are all sprockets guarded?	YES	NO	10	
4.	Are all handrails in place where needed?	YES	NO	5	
5.	Are all toeguards in place where needed?	YES	NO	5	
6.	Do you have all necessary emergency stop bottons, cables, etc?	YES	NO	15	
7.	Are emergency stops on all machinery, etc.?	YES	NO	15	
8.	Are guards being replaced after cleaning maintenance, etc.?	YES	NO	15	
9.	Have moving parts on all machinery been analyzed for guarding purposes?	YES	NO	10	
10.	Are all augers guarded?	YES	NO	10	
11.	Are all open pits, manholes, etc. guarded?	YES	NO	10	
12.	Are all trailers jacked and chocked?	YES	NO	10	
13.	Are all extended shafts cut off to specification or properly guarded?	YES	NO	10	
	Section Total				

Medical Community	Answer		Total Points	Score
1. Has a designated representative of the Company met with the physicians in the community?	YES	NO	15	
2. Are you meeting with the attending physician in individual cases?	YES	NO	10	
3. Have you opened a line of communication with area hospitals, physicians, and medical community?	YES	NO	10	
4. Are you providing the attending physician with a letter completely describing the Restricted Duty position available on work related injuries?	YES	NO	20	
5. Have you invited the medical community to tour your facility?	YES	NO	10	
6. Are you following up on a regular basis with the attending physician on time loss cases?	YES	NO	15	
7. Are you visiting employees in the hospital?	YES	NO	15	
8. Are you visiting employees on time loss at their home?	YES	NO	10	
9. Are you explaining Workman's Compensation benefits to injured employees?	YES	NO	5	
10. Are you explaining the Workman's Compensation billing procedures to the physician's office, hospital, etc.?	YES	NO	10	
Section Total			**125**	

Testing	Answer		Total Points	Score
1. Do you have all necessary monitoring equipment? (i.e confined space entry, air monitoring, etc.)	YES	NO	5	
2. Do you have written programs and procedure?	YES	NO	5	
3. Do you have specific chemical monitoring procedures?	YES	NO	5	
4. Do you have a radiation monitor and procedure?	YES	NO	5	
5. Are you in compliance with the confined space standard?	YES	NO	5	
6. Are confined spaces tested prior to entry?	YES	NO	5	
7. Do you have a H_2S monitor and procedure?	YES	NO	5	
Section Total			**35**	

Evaluation of Program Efficiency	Answer		Total Points	Score
1. Are you performing this audit at least quarterly?	YES	NO	10	
2. Are you communicating the information generated by this audit to your management team?	YES	NO	10	
3. Are you progressing on the items identified by this audit as being deficient?	YES	NO	10	
Section Total			**30**	

Reporting	Answer		Total Points	Score
1. Are you notifying the safety department immediately on all serious injuries and fatalities? Who is ressponsible for contracting OSHA within 8 hours of a fatality ? Is your legal department notified on fatalities ?	YES	NO	10	
2. Are you notifying the President on all serious injuries and fatalities?	YES	NO	10	
3. Are you notifying Worker's Compensation department or administrator on all petitions for denial of claim? Notifying on all claims ? Are claims tracked or followed ?	YES	NO	10	
4. Are you notifying Risk Management or Insurance Company on all property losses?	YES	NO	10	
5. Are you notifying Corporate Safety on all fires?	YES	NO	10	
6. Are you notifying Corporate Legal whenever any governmental agency arrives at your plant?	YES	NO	10	
7. Are you notifying Risk Management whenever the fire system is shut down?	YES	NO	10	
Section Total			**70**	

Recordkeeping	Answer		Total Points	Score
1. Is your plant logging all injuries/illnesses on the OSHA 200 log?	YES	NO	10	
2. Is your nursing staff thoroughly trained in the completion of the OSHA 200 form?	YES	NO	10	
3. Are all OSHA 200 forms being kept permanently on file in your plant?	YES	NO	10	
4. Does your management team know the OSHA criteria for recordability?	YES	NO	10	
5. Does your management team understand the recordkeeping requirements for occupational injuries and illnesses?	YES	NO	10	
6. Does your plant nurse, safety coordinator, personnel manager, and other members of the management team review each case before judging the case to be/not to be recordable?	YES	NO	10	
7. Do you possess the guide to recordkeeping requirements for occupational injuries and illnesses (April 1986)?	YES	NO	10	
8. Is your nursing staff thoroughly trained in the accurate completion of the month-end injury/illness summary report?	YES	NO	10	
9. Is the month-end injury/illness summary report being reviewed by the management team before forwarding to corporate safety?	YES	NO	10	
10. Are the nurses completing the worksheet attached to the month-end injury/illness summary report properly?	YES	NO	10	
11. Are the nurses knowledgeable in the procedure for properly calculating the loss time days?	YES	NO	10	
12. Do you have a system for tracking your lost time cases and lost time days?	YES	NO	10	
13. Does your medical staff know the criteria for a lost time case?	YES	NO	10	
14. Is your medical staff knowledgeable in the proper placement and procedure for recording petitions for denial of claims on the month-end injury/illness summary report?	YES	NO	10	
15. Is your medical staff in full knowledge of the report deadlines?	YES	NO	10	
16. Are you ensuring all claims for each month are recorded on the month-end injury/illness summary?	YES	NO	10	
17. Are all first reports and Accident Investigation reports (or photocopies) for all claims being attached to the month-end injury/illness summary?	YES	NO	10	
18. Are you checking each Accident Investigation report for completion in a timely manner?	YES	NO	10	
19. Are you attaching computer analysis of all injury/illness to your month-end report?	YES	NO	10	
20. Do you have a light duty/restricted duty log and/or tracking system?	YES	NO	10	
21. Are you insuring all first reports forms necessary are being completed?	YES	NO	10	
22. Are you insuring all necessary first reports are being completed for all necessary cases which have been evaluated by the in-plant physician?	YES	NO	10	
23. Are you insuring that the Annual (year-end) Summary Report is being completed properly and posted during the month of February?	YES	NO	10	
24. Is your medical staff knowledgeable in the procedure for handling the medical files for terminated employees? Requests by current employees ?	YES	NO	10	
25. Do you have a system for documenting your in-plant physicians evaluation?	YES	NO	10	
26. Are you sending a light duty/restricted duty letter to the treating physician on all time loss cases?	YES	NO	10	
27. Is your medical staff knowledgeable in the procedure when an employee requests access to his/her medical file?	YES	NO	10	
28. Are you utilizing outside investigators for potential fraud cases?	YES	NO	10	
29. Are your providing your management team with a weekly lost time and restricted duty list?	YES	NO	10	
30. Are all first reports and Accident Investigation reports being properly maintained and secured?	YES	NO	10	
31. Is the safety manager's developing a monthly progress report for the management team ?	YES	NO	10	
Section Total			**310**	

Safety Incentive Program	Answer		Total Points	Score
1. Do you possess a Safety Incentive Program ?	YES	NO	10	
2. Does your management team fully understand the program?	YES	NO	10	
3. Is your safety incentive program (if applicable) in compliance with this guideline?	YES	NO	10	
Section Total			30	

Supervisor's Daily Inspection Program	Answer		Total Points	Score
1. Have all supervisors been educated and thoroughly trained in this program?	YES	NO	10	
2. Are all supervisors in compliance with this program?	YES	NO	10	
3. Is the documentation for this program being properly maintained and stored?	YES	NO	10	
4. Are supervisors identifying and disciplining employees not wearing the required personal protective equipment?	YES	NO	10	
5. Are the supervisor's daily inspection forms being turned into the safety representative on a daily basis?	YES	NO	10	
6. Is the safety representative reviewing and evaluating all daily safety inspection reports?	YES	NO	10	
7. Is the plant manager reviewing and evaluating any questionable daily safety reports?	YES	NO	10	
Section Total			70	

Fall Protection Program	Answer		Total Points	Score
1. Do you have a written fall protection program?	YES	NO	10	
2. Do you have all of the necessary OSHA/ANSI approved fall protection equipment necessary for your plant?	YES	NO	10	
3. Do you have the fall protection program videotape?	YES	NO	5	
4. Are all appropriate employees properly trained in the use of fall protection equipment?	YES	NO	10	
5. Are all employees required to utilize the fall protection equipment using this equipment properly?	YES	NO	10	
6. Do you have a documented inspection procedure for your fall protection equipment?	YES	NO	10	
7. Do you have a signed document in each employee's personnel file showing he/she has read and understands the fall protection program?	YES	NO	10	
8. Do you have the fall protection program on file?	YES	NO	10	
9. Is your fall protection program in compliance?	YES	NO	10	
10. Do you have all of the necessary tie-off points?	YES	NO	10	
SECTION TOTAL			95	

Confined Space Entry Procedure	Answer		Total Points	Score
1. Do you have a written confined space program ?	YES	NO	10	
2. Do you have the necessary self contained breathing apparatus(s)?	YES	NO	10	
3. Are the self contained breathing apparatus inspected (documented) on a weekly basis?	YES	NO	10	
4. Are all appropriate employees properly trained in the use of the self contained breathing apparatus?	YES	NO	10	
5. Are all employees aware of and fully understand the safety procedures for blood tankers? boilers ? other vessels ?	YES	NO	25	
6. Have all appropriate employees signed documentation stating they have read and fully understand this policy/procedure?	YES	NO	10	
7. Do you have an oxygen monitor?	YES	NO	10	
8. Do you have other required monitoring devices?	YES	NO	10	

	Answer		Total Points	Score
9. Are the above instruments calibrated?	YES	NO	10	
10. Is the calibration documented?	YES	NO	10	
11. Are all appropriate employees properly trained in the use of the oxygen monitor and other detectors?	YES	NO	10	
12. Do you have the lifeline system in your plant?	YES	NO	10	
13. Is the lifeline system being utilized?	YES	NO	10	
14. Do you utilize spark proof flashlights, tools, etc., when working in the silos?	YES	NO	10	
15. Is your confined space entry policy/procedure being utilized and enforced?	YES	NO	25	
16. Is your confined space entry policy/procedure on file?	YES	NO	10	
17. Are all appropriate employees and your management team properly trained in confined space entry procedures?	YES	NO	25	
Section Total			**215**	
Subcontractor's Policy	**Answer**		**Total Points**	**Score**
1. Do you have a subcontractor's safety policy?	YES	NO	10	
2. Are your management team members aware of and fully understand this policy?	YES	NO	10	
3. Is a copy of this policy included in your Hazard Communication program?	YES	NO	10	
4. Are all appropriate subcontractors provided a copy of this policy and the attached questionnaire?	YES	NO	10	
5. Are all subcontractors following the company's safety policies/ procedures?	YES	NO	25	
6. Does the safety manager inspect all subcontractor's working on premises on at least a weekly basis?	YES	NO	10	
7. Are items cited during this inspection documented and provided to the subcontractor and appropriate company departments?	YES	NO	25	
8. Are the questionnaires being completed by the subcontractors?	YES	NO	10	
9. Are the deficient items identified being corrected in a timely manner?	YES	NO	20	
Section Total			**120**	
Radiation Procedures	**Answer**		**Total Points**	**Score**
1. Do you have a written radiation safety program?	YES	NO	10	
2. Do you have a written safety procedures?	YES	NO	25	
3. Is your program in compliance?	YES	NO	25	
4. Do you have a radiation monitor?	YES	NO	10	
5. Is your radiation monitor properly calibrated?	YES	NO	10	
6. Are documented inspections conducted on at least a weekly basis?	YES	NO	25	
7. Are all employees properly trained?	YES	NO	10	
8. Is your unit being sent back to the manufacturer for repairs and calibration?	YES	NO	25	
9. Are annual inspections conducted by the manufacturer?	YES	NO	10	
10. Is your unit registered with your appropriate state agency?	YES	NO	10	
11. Do you have the appropriate posting placed in your bulletin board?	YES	NO	10	
12. Are the operator instructions included in your written program?	YES	NO	10	
13. Are all employees performing the inspection and use the radiation monitor properly trained?	YES	NO	10	
14. Is your safety program in compliance with your state regulations?				
Section Total			**215**	
Hazard Communication Program	**Answer**		**Total Points**	**Score**
1. Do you have a written Hazard Communication program?	YES	NO	25	
2. Is a complete list of chemicals for your plant included in your written program?	YES	NO	10	

3. Is your routine and non-routine training outlined in your written program?	YES	NO	10	
4. Do you have the MSDS reports for all chemicals noted in your list of chemicals?	YES	NO	10	
5. Do you have a letter requesting the MSDS for any chemicals which you do not possess an MSDS report?	YES	NO	10	
6. Do you possess the hazard communication training video tapes?	YES	NO	5	
7. Have all appropriate employees been properly trained in the procedures outlined in your hazard communication program?	YES	NO	25	
8. Is your hazard communication notice posted?	YES	NO	10	
9. Are all necessary management team members provided a copy of the hazard communication program?	YES	NO	10	
10. Are emergency procedures included in your hazard communication program?	YES	NO	10	
11. Do you have a chemical spill procedure?	YES	NO	10	
12. Do you have a chemical spill cart or station?	YES	NO	10	
13. Have all appropriate employees signed the documentation sheet stating that they have read and understand this policy/procedure?	YES	NO	10	
14. Does your state require a copy of your hazard communication program be kept on file with the local fire department, disaster preparedness agency, etc.?	YES	NO	10	
15. If yes in #14, have these agencies been provided a copy of your program to be kept on file with the local fire department, disaster preparedness agency, etc.?	YES	NO	10	
16. Do the subcontractors working on premises have a hazard communication program?	YES	NO	10	
17. Have all subcontractors submitted a list of all chemicals plus MSDS reports for each chemical to the safety manager before beginning a project?	YES	NO	10	
18. Does the purchasing department, storeroom, maintenance, and other applicable departments require MSDS reports for all new chemicals before allowing their use in the plant?	YES	NO	10	
19. Are subcontractor's chemicals inspected to insure compliance?	YES	NO	10	
20. Is your hazard communication program in compliance with OSHA regulations?	YES	NO	25	
Section Total			240	

Safety Equipment Procedure	Answer		Total Points	Score
1. Have all employees read, understand, and signed your PPE policies? Do you possess a written PPE program complying with all elements of the April 5, 1994 Final Ruling ?	YES	NO	10	
2. Do all new employees read, understand, and sign this policy? Has someone evaluated and "certified" the PPE for each job ? Is your initial and on-goinig training documented ? Are employees instructed on how to wear and care for PPE ? Instructed on how to replace worn or broken PPE ?	YES	NO	10	
3. Is this documentation kept in the employee's personnel file? Are training records maintained in the written program ?	YES	NO	10	
4. Is this policy posted? Is the written program readily accessible ?	YES	NO	10	
5. Are all employees disciplined in accordance with this policy? Is this disciplinary action documented ?	YES	NO	10	
6. Are daily safety inspections being performed and documented by all supervisors?	YES	NO	10	
Section Total			60	

Unsafe Acts Procedures	Answer		Total Points	Score
1. Have all employees read, understand and signed this procedure?	YES	NO	10	
2. Is this policy posted?	YES	NO	10	
3. Are all employees disciplined in accordance with this policy? Is this discipline documented ?	YES	NO	10	
Section Total			30	

	Seat Belt Policy	Answer		Total Points	Score
1.	Do you have a seat belt program?	YES	NO	10	
2.	Have all management team members read and understand this policy?	YES	NO	10	
3.	Are the seat belt stickers in all company vehicles on location?	YES	NO	10	
4.	Are the seat belt signs posted at the exits from your plant?	YES	NO	10	
	Section Total				40

	Light Duty Policy	Answer		Total Points	Score
1.	Do you have a light or restricted duty policy?	YES	NO	10	
2.	Have all management team members read and understand this policy?	YES	NO	10	
3.	Is your plant in compliance with this policy?	YES	NO	10	
4.	Is the safety manager developing and distributing a weekly restricted duty report?	YES	NO	10	
	Section Total			40	

	Reporting of Accident Policy	Answer		Total Points	Score
1.	Do you have a accident reporting policy?	YES	NO	10	
2.	Is a copy of this policy posted?	YES	NO	10	
3.	Are all employees in full understanding of this policy?	YES	NO	10	
4.	Is your plant in compliance with this policy?	YES	NO	10	
	Section Total			40	

	Forklift Operator's Certification Program	Answer		Total Points	Score
1.	Do you possess a written forklift operator's program	YES	NO	10	
2.	Are all appropriate employee's properly trained in this program?	YES	NO	10	
3.	Is classroom instruction and hands-on instruction included in your training program?	YES	NO	10	
4.	Is written testing and hands-on testing conducted in your training program?	YES	NO	10	
5.	Is your program in compliance with OSHA standards?	YES	NO	10	
6.	Is your program in compliance with the Company's standards?	YES	NO	10	
7.	Are certification cards or other identification awarded to certified operators?	YES	NO	10	
8.	Are safety videotapes used on the training program?	YES	NO	10	
9.	Are vehicle inspection techniques taught in the training program?	YES	NO	10	
	Section Total			90	

	Safety Glass Policy	Answer		Total Points	Score
1.	Do you possess a safety glass policy? Do you possess a written program or section within your written PPE program ?	YES	NO	10	
2.	Are you utilizing X type of safety glasses?	YES	NO	10	
3.	Do you have a plant prescription safety glass policy?	YES	NO	10	
4.	Have your identified the areas in which safety glasses are required in your plant?	YES	NO	10	
5.	have all employees been educated and trained in the safety glasses requirements in your plant? Are employees trained on how to wear and care for safety glasses ? Are employees trained on where and when to replace worn or broken safety glasses ?	YES	NO	10	
6.	Is your plant safety glass requirement posted?	YES	NO	10	
	Section Total			60	

Head Protection Program	Answer		Total Points	Score
1. Do you possess possess head protection program? Is this a written program or part of your written PPE program ?	YES	NO	10	
2. Is your plant utilizing approved hard hats? Certified ?	YES	NO	10	
3. Are all personnel wearing the hard hats properly? Are employees properly trained on wearing and caring for their hard hat ? Trained on when to replace worn or broken PPE ?	YES	NO	10	
4. Have all personnel been trained in the proper use and care of hard hats? Is the issue of hard hatr stickers addressed ?	YES	NO	10	
5. Are periodic inspections being conducted on the hard hats?	YES	NO	10	
6. Is your head protection program in compliance with OSHA and the company's requirements?	YES	NO	10	
Section Total			60	

Cumulative Trauma Prevention Program (Ergonomics)	Answer		Total Points	Score
1. Have you analyzed each job for possible ergonomic improvements? Do you possess a copy of the Red Meat Guidelines on Ergonomics ? Are you prepared to comply with the new OSHA General Industry standard scheduled for September, 1994?	YES	NO	20	
2. Do you possess a written cumulative trauma prevention program?	YES	NO	10	
3. Have you applied any of the ergonomic study recommendations in your plant?	YES	NO	20	
4. Do you possess the alcohol and controlled substance program in your plant?	YES	NO	20	
5. Is the alcohol and controlled substance testing equipment calibrated and functioning properly?	YES	NO	10	
6. Are all appropriate personnel properly trained in the use of alcohol and controlled substance testing equipment?	YES	NO	10	
7. Do you possess the OSHA heat/cold stress information?	YES	NO	10	
8. Have you developed a heat/cold stress film and program?	YES	NO	10	
9. Have you evaluated/analyzed all hot and cold areas in your plant for safety purposes?	YES	NO	10	
10. Have the appropriate preventative measures been taken to protect employees working in hot or cold environments?	YES	NO	10	
11. Have you acquire the recent research on ergonomics?	YES	NO	10	
12. Do you possess a copy of the proposed OSHA standard ?	YES	NO	10	
13. Are you utilizing conservative treatment ?	YES	NO	10	
14. Are you utilizing a hand exercise program?	YES	NO	10	
15. Have you tested the pre-work exercise program?	YES	NO	10	
16. Have you developed a cumulative trauma prevention education program for your management team?	YES	NO	10	
17. Have you developed a cumulative trauma prevention education program for your hourly workforce?	YES	NO	10	
18. Have you analyzed your facility for possible job rotation and/or job combination positions?	YES	NO	10	
19. Have you reviewed your alternate duty, restricted duty, and job change program?	YES	NO	10	
20. Have you tested alternative equipment?	YES	NO	10	
21. Have you reviewed the literature regarding ergonomic studies?	YES	NO	10	
22. Have you developed a "hardening" exercise program?	YES	NO	10	
23. Have you reviewed the individual equipment studies?	YES	NO	10	
24. Do you possess appropriate ergonomic literature and reference books on location?	YES	NO	5	
25. Do you possess a written ergonomic program?	YES	NO	10	
26. Are you properly managing employees with cumulative trauma illnesses?	YES	NO	10	
27. Has your plant implemented a program to manage employees with cumulative trauma illnesses?	YES	NO	10	

28.	Has your management team provided input into the development of the cumulative trauma questionnaire?	YES	NO	5	
29.	Have all safety committees and communication committees completed the non-supervisory training program?	YES	NO	10	
30.	Are all appropriate hourly personnel utilizing a near miss program for ergonomics?	YES	NO	10	
	Section Total			**320**	

	Safe Lifting Program	Answer		Total Points	Score
1.	Do you possess a written safe lifting program?	YES	NO	20	
2.	Have you reviewed your selection, physical examination, orientation, and placement procedures to insure compliance with the company standards?	YES	NO	20	
3.	Have you conducted a Job Safety Analysis (JSA) on all jobs addressing ergonomics and safe lifting considerations?	YES	NO	10	
4.	Do you have a pre-employment safe lifting program?	YES	NO	10	
5.	Do you have a follow-up safe lifting program for all appropriate employees during the probationary period?	YES	NO	10	
6.	Do you have annual safe lifting training for all appropriate personnel?	YES	NO	10	
7.	Are all safe lifting training sessions documented?	YES	NO	10	
8.	Does your safe lifting program meet all company requirements?	YES	NO	10	
9.	Are you utilizing safe lifting promotional posters, etc.?	YES	NO	10	
10.	Are you utilizing "weight lifter" belts, or other types of back supports for appropriate personnel? Are your employees trained ?	YES	NO	10	
11.	Are you conducting a thorough investigation on all alleged back injuries?	YES	NO	10	
12.	Are you utilizing the insurer's investigation services to assist in the investigation of alleged back injuries or other questionable type of alleged injuries?	YES	NO	10	
	Section Total			**140**	

	Bloodborne Pathogen Program	Answer		Total Points	Score
1.	Have all employees read, understand and signed this procedure? Do you possess a written bloodborne pathogen program in compliance with the OSHA standard ?	YES	NO	10	
2.	Is this procedure posted? Are all employees trained ? Is this training documented ? Are appropriate employees provided Hepatitis B injections ? Do you possess all of the appropriate PPE ? Sharps containers ?	YES	NO	10	
3.	Are all employees disciplined in accordance with this policy/procedure? Do you [possess the appropriate clean-up and disposal procedures for biohazardous materials ?	YES	NO	10	
	Section Total			**30**	

	Control of Hazardous Energy	Answer		Total Points	Score
1.	Do you have written lockout/tagout program ? Is this program in compliance with the OSHA standard ?	YES	NO	10	
2.	Have all management team members read and understand this program? Have all employees been trained in this program ? Is this training documented ?	YES	NO	10	
3.	Do you possess all of the required equipment for this program ? Has all electrical, pneumatic, hydraulic, steam , and other energy sources been analyzed and appropriate control mechanisms been installed ? Has stored energy issues been addressed ?	YES	NO	10	
4.	Are employees disciplined for not complying with this program ? Is the disciplinary action documented ?	YES	NO	10	
	Section Total				**40**

New standards, such as Bloodborne Pathogens, will require questions to be developed and added to the structure of the audit mechanism.

Indoor Air Quality	Answer		Total Points	Score
1. Do you have a copy of the proposed OSHA standard? Have you addressed air quality issues ? Have you conducted air quality sampling ?	YES	NO	10	
2. Have you discussed this proposed standard with your management team ?	YES	NO	10	
3. Are you taking a proactive approach to this proposed standard?	YES	NO	10	
4. Do you have any special air quality issues or areas in your facility?	YES	NO	10	
Section Total			40	

Workplace Violence	Answer		Total Points	Score
1. Do you possess any "at risk" positions ? Personnel ? Medical ? President's office ? (Note: OSHA cites this area under the general duty clause - 5(A)(1)).	YES	NO	10	
2. Have you addressed potential security risk areas ? Parking lots ?	YES	NO	10	
3. Have you addressed in house security issues?	YES	NO	10	
4. Are employees "at risk" for workplace violence ?	YES	NO	10	
Section Total			40	

TOTAL POINTS SCORED DIVIDED BY THE TOTAL POIINTS POSSIBLE WILL PROVIDE A PERCENTAGE EFFICIENCY WITH YOUR SAFETY AND HEALTH PROGRAM. YOU MAY ADD ADDITIONAL SAFETY AND HEALTH CONCERNS OR NEW STANDARDS AS NECESSARY.				
May also want to consider adding sections				
on robotics, laser safety, tuberculosis, and				
any other applicable safety and health situation/standard.				

APPENDIX C

OSHA INSPECTION CHECKLIST

The following is a recommended checklist for Safety and Resource Control Professionals in order to prepare for an OSHA inspection:

1. Assemble a team from the management group and identify specific responsibilities in writing for each team member. The team members should be given appropriate training and education and should include, but not be limited to:
   a. An OSHA inspection team coordinator
   b. A document control individual
   c. Individuals to accompany the OSHA inspector
   d. A media coordinator
   e. An accident investigation team leader (where applicable)
   f. A notification person
   g. A legal advisor (where applicable)
   h. A law enforcement coordinator (where applicable)
   i. A photographer
   j. An industrial hygienist
2. Decide on and develop a company policy and procedures to provide guidance to the OSHA inspection team.
3. Prepare an OSHA inspection kit, including all equipment nec-

essary to properly document all phases of the inspection. The kit should include equipment such as a camera (with extra film and batteries), a tape player (with extra batteries), a video camera, pads, pens, and other appropriate testing and sampling equipment (e.g., a noise level meter, an air sampling kit).

4. Prepare basic forms to be used by the inspection team members both during and after the inspection.
5. When notified that an OSHA inspector has arrived, assemble the team members along with the inspection kit.
6. Identify the inspector. Check his or her credentials and determine the reason for and type of inspection to be conducted.
7. Confirm the reason for the inspection with the inspector (targeted, routine inspection, accident, or in response to a complaint?)
   a. For a random or target inspection:
      - Did the inspector check the OSHA 200 Form?
      - Was a warrant required?
   b. For an employee complaint inspection:
      - Did inspector have a copy of the complaint? If so, obtain a copy.
      - Do allegations in the complaint describe an OSHA violation?
      - Was a warrant required?
      - Was the inspection protested in writing?
   c. For an accident investigation inspection:
      - How was OSHA notified of the accident?
      - Was a warrant required?
      - Was the inspection limited to the accident location?
   d. If a warrant is presented:
      - Were the terms of the warrant reviewed by local counsel?
      - Did the inspector follow the terms of the warrant?
      - Was a copy of the warrant acquired?
      - Was the inspection protested in writing?
8. The Opening conference
   a. Who was present?

b. What was said?
c. Was the conference taped or otherwise documented?

9. Records
   a. What records were requested by the inspector?
   b. Did the document control coordinator number the photocopies of the documents provided to the inspector?
   c. Did the document control coordinator maintain a list of all photocopies provided to the inspector?
10. Facility Inspection
   a. What areas of the facility were inspected?
   b. What equipment was inspected?
   c. Which employees were interviewed?
   d. Who was the employee or union representative present during the inspection?
   e. Were all the remarks made by the inspector documented?
   f. Did the inspector take photographs?
   g. Did a team member take similar photographs?[1]

ENDNOTE

OSHA Criminal Referrals

1. Schneid, T., *Preparing for an OSHA Inspection*, Kentucky Manufacturer, February 1992.

APPENDIX D

EMPLOYMENT-RELATED LAWS

Safety and resource control professionals should be aware that there are a myriad of laws in the employment arena that can create possible liability if not properly addressed. Recently I was asked if there is one location where virtually all the employment-related laws could be located, and I could not provide a good answer without referencing the individual to several texts. The following summary of many of the federal and state laws that could impact most organizations represents an attempt to resolve this issue.

POSSIBLE FEDERAL LAWS APPLICABLE TO EMPLOYMENT

1. Americans With Disabilities Act (42 U.S.C. §12101)
   - Applies to organizations with 15 or more employees.
   - Prohibits discrimination in all employment areas (Title I) against qualified individuals with disabilities.
   - Requires reasonable accommodation.
   - Acquire information from the Equal Employment Opportunity Commission (EEOC).
2. Civil Rights Act of 1866 (42 U.S.C. §1981)

- No size requirement.
- Applies to public and private sector.
- Applies to all persons under recent U.S. Supreme Court decisions.
- Religious bias not covered.
- See *Patterson v. McLean Credit Union*, 109 S.Ct. 2263 (1989), but note that the Civil Rights Act of 1991 restored much of the protection restricted by this case.

3. Civil Rights Act of 1964 (42 U.S.C. §2000e)
   - 15 or more employees.
   - Bar discrimination based on race, sex, color, religion, or national origin.
   - Sex discrimination includes sexual harassment in employment.
4. Civil Rights Act of 1991
   - Negates several Supreme Court decisions regarding the Civil Rights Act of 1964 and expanded remedies in sex discrimination and sexual harassment cases.
   - Amended Civil rights act of 1866 to include prohibitions of race discrimination in the contract area.
   - Permits injunctive relief, declaratory relief, and attorney fees in mixed motive cases.
   - Applies to private and public sector.
   - Cap on damages depending on the size of the employer.
5. Age discrimination in Employment Act (29 U.S.C. §621)
   - 20 or more employees.
   - Applies to public and private sector.
   - Prohibits discrimination based on age. Protects individuals over the age of 40.
   - Age cap of 70 years of age removed.
6. Drug Free Workplace Act of 1988 (41 U.S.C. §701, §5151–5160)
   - Applies to federal contractors and grantees.
   - Employers must certify a drug free workplace.
   - Employees must report drug convictions.

7. Consolidated Omnibus Budget Reconciliated Act of 1985 (COBRA) (26 U.S.C. §162)
   - Applies to employee benefit plan.
   - Mandates coverage of group health benefits when employee voluntarily or involuntarily terminates employment.
   - Requires employers make available group health care coverage to terminated employee and their dependant for a minimum of 18 months.
8. Fair Labor Standards Act (29 U.S.C. §201)
   - Requires payment of minimum wage.
   - No discrimination on the basis of sex in payment of wages.
   - Time and a half for work over 40 hours.
9. Employee Polygraph Protection Act of 1988 (29 U.S.C. §2001)
   - Applies to public and private sector.
   - Prohibits most employers from using polygraph examinations in most pre-employment situations, except for security or related businesses. Current employees can only be tested during ongoing investigation.
   - Numerous restrictions and prohibitions.
10. Employee Retirement Income Security Act of 1974 (ERISA) (29 U.S.C. §1001)
    - Applies to any employee benefit plan.
    - Tax, reporting, administration, funding, and other requirements.
11. Equal Pay Act of 1963 (29 U.S.C. §206)
    - Applies to any employer with 2 or more employees.
    - Requires equal pay for male and female workers with similar skills, effort, responsibilities, etc.
12. Family and Medical Leave Act of 1993 (29 U.S.C. §2601)
    - 50 or more employees.
    - Eligible to up to 12 weeks unpaid leave for serious health condition of the employee or family member or birth or adoption of child.
13. Immigration Reform and Control Act of 1986 (8 U.S.C. §1324a)

- All employers must verify and document eligibility to work in the United States (i.e., 1–9).
- Civil and criminal penalties for hiring illegal aliens.
- Unfair employment practice for employers (4 or more employees) to discriminate on the basis of national origin or citizenship status.

14. Labor–Management Relations Act of 1947 (29 U.S.C. §141)
    - Allows employees to file a court action against the employer for breach of contract or against a union for breach of duty of fair representation.
15. Occupational Safety and Health Act of 1970 (29 U.S.C. §651)
    - Private sector—1 or more employees; public sector—usually by state law.
    - Requires maintenance of safe and healthful work environment.
    - Civil penalties up to $70,000.00 per violation.
    - Criminal penalties.
16. Veterans Reemployment Rights Act of 1974 (38 U.S.C. §2021)
    - Applies to public and private sector.
    - Active military and guard training entitled to guaranteed job security.
    - Other protections.
17. Vietnam Era Veterans Readjustment Assistance Act of 1974 (38 U.S.C. §2012) and Employment and Training of Veterans Act (39 U.S.C. §4211)
    - Federal projects over $10,000.00.
    - Requires written affirmative action plans.
18. Vocational Rehabilitation Act of 1973 (29 U.S.C. §701)
    - Federal contracts over $10,000.00.
    - Affirmative action plan required for hiring and advancing qualified individuals with disabilities.
19. Worker Adjustment and Retraining Notification (WARN) (29 U.S.C. §2101)
    - 100 or more employees.
    - Prohibits ordering of plant closing and mass layoffs without 60 days' advance notice.

The Pregnancy Discrimination Act [42 U.S.C. §2000 e(k)]

- Prohibits sexual discrimination and amends the Civil Rights Act of 1964 to include pregnancy, childbirth, and pregnancy related medical conditions as protected against employment discrimination.

These are the major federal laws that could impact most organizations. There are a similar number of federal laws governing the labor relations area as well as a substantial number of state laws (i.e., child labor, workers' compensation, unemployment) and possibly local laws (discrimination based of sexual preference) that impact the employment area. Prudent safety and resource control professionals should assess the applicability of these laws to their organizations and ensure that their programs, policies, and procedures comply with all aspects of these laws. Knowing that the particular law applies to your organization and addressing the specific requirements of the law is essential in ensuring compliance and avoiding costly potential problems now and in the future.

APPENDIX E

STATE EMPLOYMENT-RELATED LAWS

Safety and resource control professionals should also possess a working knowledge of the applicable state and local laws. Virtually all states possess laws that mirror the federal laws and/or expand into areas over which only the state government possesses jurisdiction. Using Kentucky as an example, following is a summary of many of the areas of employment related laws which may be applicable in your state:

KENTUCKY SPECIFIC LAW

(Kentucky Revised Statutes, or KRS)

- KRS 334.040 forbids discrimination in employment on the basis of race, color, religion, national origin, sex, or age.
- KRS 207.130 prohibits discrimination in employment because of handicap.
- KRS 344.030 forbids discrimination based on pregnancy.
- Kentucky Public Acts, House Bill 23-X (L. 1976) prohibits an employer from attempting to deprive an employee of his or her employment because the employee receives a summons, serves as a juror, or attends court for prospective jury service.

- KRS 433.310 forbids anyone from inducing or enticing any person, who has contracted his or her employer for a specific period of time, to leave the service of the employer during the term without the employers consent.
- KRS 121.310 forbids employers from interfering with the voting rights of their employees.
- Kentucky Public Acts, H.B. 100 (1978) and KRS 335(B)(2)(1) provides that no person shall be disqualified from public employment, nor shall a person be disqualified from pursuing, practicing, or engaging in any occupation solely because of a prior conviction of a crime, unless the crime is limited to convictions for felonies, or high misdemeanors or involves moral turpitude.
- KRS 38.460 makes it unlawful for any person to wilfully deprive a member of the National Guard of his or her employment.
- KRS 337.423 makes it unlawful to discriminate in the payment of wages on the basis of sex.
- KRS 339.205 addresses child labor issues. This Act bars employment of minors under the age of 14 and regulates employment for minors ages 14–18.
- KRS 338.011 addresses occupational safety and health for public and private sector employers.
- KRS 341.005 establishes and regulates unemployment insurance.
- KRS 337.010 establishes and regulates wages and hours of work in Kentucky. This Act establishes the minimum wage, restrictions on deductions, and other aspects of compensation.
- KRS 342.0011 establishes and regulates workers' compensation. (In Kentucky, volunteer firefighters are specifically included in coverage under the Act.)
- KRS 337.420 establishes and regulates equal pay and gender-based discrimination in pay.

As can be seen, several of the state statutes mirror the federal employment laws however several of these laws are specific state laws. Prudent safety and resource control professionals should become familiar not only with the federal employment related laws but the state employment related laws and laws specific to their industry (i.e., civil service, workers compensation) in order to avoid potential problems and possible liabilities. Information regarding your state specific laws can normally be acquired from your local legal counsel or nearest law library. Virtually every state possesses laws governing the employment area, and the breadth and scope of the laws may be significantly different. Knowledge of these laws is the first step in ensuring compliance and avoiding potential liability.

APPENDIX F

OSHA PUBLICATIONS

100 Most Frequently Cited OSHA Construction Standards in 1991: A Guide for the Abatement of the Top 25 Associated Physical Hazards	OSHA 1995 Reprint: March 1995
A Guide to Scaffold Use in the Construction Industry	OSHA 3150 Revised: 1998
All About OSHA	OSHA 2056 Revised: 1995
Analysis of Construction Fatalities—The OSHA DataBase 1985–1989	OSHA 1990 Printed: November 1990
Asbestos Standard for General Industry	OSHA 3095 Revised: 1995
Asbestos Standard for the Construction Industry	OSHA 3096 Revised: 1995
Assessing the Need for Personal Protective Equipment: A Guide for Small Business Employers	OSHA 3151 Printed: 1997
Bloodborne Pathogens & Acute Care Facilities	OSHA 3128 Printed: 1992

Bloodborne Pathogens & Long-Term Care Workers	OSHA 3131 Printed: 1992
Chemical Hazard Communication	OSHA 3084 Revised: 1998
Cold Stress Card *English Version*	OSHA 3156 English Version Revised: 1999
Cold Stress Card *Spanish Version*	OSHA 3158 Spanish Version Revised: 1999
Concepts & Techniques of Machine Safeguarding	OSHA 3067 Revised: 1992
Concrete & Masonry Construction	OSHA 3106 Revised: 1998
Construction Industry Digest	OSHA 2202 Revised: 1998
Construction Resource Manual	WEBSITE Revised: 1998
Consultation Services for the Employer	OSHA 3047 Revised: 1997
Control of Hazardous Energy (Lockout/Tagout)	OSHA 3120 Revised: 1997
Controlling Electrical Hazards	OSHA 3075 Revised: 1997
Controlling Occupational Exposure to Bloodborne Pathogens in Dentistry	OSHA 3129 Printed: 1992
Crane or Derrick Suspended Personal Platforms	OSHA 3100 Revised: 1993
Employee Workplace Rights	OSHA 3021 Revised: 1997
Employer's Guide to Teen Worker Safety	DOL Web Presentation
Employers Rights and Responsibilities Following an OSHA Inspection	OSHA 3000 Revised: 1999

Ergonomics Program Management Guidelines for Meatpacking Plants	OSHA 3123 Reprinted: 1993
Ergonomics: The Study of Work	OSHA 3125 Printed: 1991
Excavations	OSHA 2226 Revised: 1995
Fall Protection in Construction	OSHA 3146 Revised: 1998
Federal Agency Poster	OSHA POSTER
Federal Employer Rights & Responsibilities Following an OSHA Inspection	OSHA Web Presentation 1996
Grain Handling	OSHA 3103 Revised: 1996
Ground Fault Protection on Construction Sites	OSHA 3007 Revised: 1998
Guidelines For Preventing Workplace Violence for Health Care and Social Service Workers	OSHA 3148 Revised: 1996
Hand and Power Tools	OSHA 3080 Revised: 1998
Handbook for Small Businesses	OSHA 2209 Revised: 1996
Hazard Communication Guidelines for Compliance	OSHA 3111 Revised: 1998
Hazardous Waste Incinerators	GAO/RCED 95-17 Printed: January 1995
Hazardous Waste Operations and Emergency Response	OSHA 3114 Revised: 1997
Hearing Conservation	OSHA 3074 Revised: 1995
Heat Stress Card	OSHA 3154/3155 Printed: 1998

Hospitals and Community Emergency Response—What You Need to Know	OSHA 3152 Printed: 1997
How to Prepare for Workplace Emergencies	OSHA 3088 Revised: 1995
How to Prevent Needle Stick Injuries: Answers to Some Important Questions Needle Stick Safety Evaluation Form... The following is the most recent copy of the Sample "Safety Feature Evaluation Form", developed by the Training for Development of Innovative Control Technology Project (TDICT), Trauma Foundation, San Francisco, CA 94110. OSHA has permission to reprint this form in both traditional and electronic publishing formats. Since the form is copyrighted, however, extensive use of the form by others may require additional permission from the copyright holders. Other evaluation forms for different devices also are available from TDICT	OSHA Booklet Sample Form Printed: 1998
Industrial Hygiene	OSHA 3143 Revised: 1998
Job Hazard Analysis	OSHA 3071 Revised: 1998
Job Safety & Health Protection Poster	OSHA POSTER
Job Safety & Health Quarterly (Vol. 10 No. 2 Winter 1999). Job Safety & Health Quarterly (Vol. 10 No. 1 Fall 1998). Job Safety & Health Quarterly (Vol. 9 No. 4 Summer 1998). Job Safety & Health Quarterly (Vol. 9 No. 3 Spring 1998). Job Safety & Health Quarterly (Vol. 9 No. 1–2 Fall/Winter 1998). Job Safety & Health Quarterly (Vol. 8 No. 4 Summer 1997).	Subscription Information
Jobsite Safety Handbook (NAHB/OSHA)	NAHB/OSHA

Keeping You Workplace Safe (Brochure)	OSHA Brochure Flyer
Lead in Construction	OSHA 3142 Printed: 1993
OSHA Log and Summary of Occupational Injuries & Illnesses	OSHA 200 Log Form
Material Safety Data Sheet (MSDS) Form	OSHA 174 Form
Materials Handling and Storing	OSHA 2236 Revised: 1998
Methylene Chloride	OSHA 3144 Printed: 1997
Methylene Chloride—Small Entity Compliance Guides	OSHA Web Presentation
4,4′ Methylenedianiline (MDA) in the Construction Industry	OSHA 3137 Reprint: 1993
4,4′ Methylenedianiline (MDA) for General Industry	OSHA 3135 Reprint: 1993
Model Exposure Control Plan for Home Care: A Guide for Hospice/Home Agencies on the Bloodborne Pathogens Standards	OSHA Printed: 1994
Occupational Exposure to Cadmium	OSHA 3136 Printed: 1992
Occupational Exposure to Bloodborne Pathogens—Precautions for Emergency Responders	OSHA 3130 Revised: 1998
Occupational Safety & Health Guidance Manual for Hazardous Waste Site Activities:	NIOSH OSHA USCG EPA Revised: 1998
OSHA Inspections	OSHA 2098 Revised: 1998
OSHA Performance Review FY 1996	OSHA/DOL Printed: April 1997

OSHA Publications & Audiovisual Programs	OSHA 2019 Revised: 1998
OSHA Supplementary Record of Occupational Injuries & Illnesses	OSHA Form 101
Outreach Training Program	OSHA Training Revised: July 1997
Personal Protective Equipment	OSHA Web Revised: 1995
Process Safety Management Guidelines for Compliance	OSHA 3133 Reprinted: 1994
Process Safety Management	OSHA 3132 Reprinted: 1994
Recommendations for Workplace Violence Prevention Programs in Late-Night Retail Establishments	OSHA 3153 Printed: 1998
Recordkeeping Guidelines for Occupational Injuries and Illnesses	OSHA Web 1986
Respiratory Protection	OSHA 3079 Revised: 1998
Respiratory Protection Standard [Small Entity Compliance Guide for] (CFR 1910. 134) **SEE BELOW—Small Entity Compliance Guide for Respiratory Protection**	OSHA SECG Revised: 1998
Safety and Health Guide for the Chemical Industry	OSHA 3091 Printed: 1986
Safety and Health Guide for the Meatpacking Industry	OSHA 3108 Printed: 1988
Safety and Health Guide for the Microelectronics Industry	OSHA 3107 Printed: 1988
Safety and Health Standards for the Construction Industry	OSHA 3149 Printed: 1996
Selected Construction Regulations (SCOR) for the Home Building Industry (29 CFR 1926)	OSHA Web 1997

Selected Occupational Fatalities	OSHA Web
Servicing Single-Piece and Multi-Piece Rim Wheels	OSHA 3086 Revised: 1998
Shipyard Industry	OSHA 2268 Revised: 1998
Sling Safety	OSHA 3072 Revised: 1996
Small Entity Compliance Guide for OSHA's Abatement Verification Regulation (29 CFR 1903.19)	OSHA 1997
Small Entity Compliance Guide for Respiratory Protection Standard (CFR 1910.134): 328 Pages Copies of the 1910.134 and other related materials are available as separate documents. All of the following items are included in the 328 Page Guide: *1910 0134* *Appendix A* *Appendix B-1* *Appendix B-2* *Appendix C* *Appendix D* Question & Answers: *Q&A* Compliance Directive: *CPL 2-0.120* News Release: *9/24/98 Release*	Revised 1998
Special Reports	OSHA Web
Stairways and Ladders	OSHA 3124 Revised: 1997
Summary Report on OSHA Inspections Conducted at SuperFund Incinerator Sites	OSHA 1993 Printed: September, 20, 1993
Training Requirements in OSHA Standards and Training Guidelines	OSHA 2254 Revised: 1998
Underground Construction (Tunneling)	OSHA 3115 Revised: 1996

Working Safely with Video Display Terminals	OSHA 3092 Revised: 1997
Voluntary Protection Program (VPP)—So You Want to Apply to VPP	OSHA Printed: 1997
What to Expect During OSHA's Visit	OSHA Printed: 1997
You've Been Selected to Be a VPP Onsite Team Member ... What Now?	OSHA Printed: 1997

APPENDIX G

GENERAL GUIDELINES FOR EFFECTIVE MANAGEMENT OF WORKERS' COMPENSATION

Safety and resource control professionals responsible for the management of workers' compensation within the organization will find that an effective management system can control and minimize the costs related to this required administrative system while also maximizing the benefits to the injured or ill employee. Although the workers' compensation system is basically reactive in nature, safety and health professionals should develop a proactive management system through which to effectively manage the workers' compensation claims once incurred within the organization. The following basic guideline can be used to implement an effective workers' compensation management system.

1. Become completely familiar with the rules, regulations, and procedures of the workers' compensation system in your state. A mechanism should be initiated to keep the professional updated with all changes, modifications, or deletions within the workers' compensation law or regulations. A copy of these laws and rules can normally be acquired from your state's workers' compensation agency at no cost. The state bar association, universities, and law schools in many states have published texts and other publications to assist in interpreting the laws and rules.

2. A management system should be designed around the basic management principles of Planning, Organizing, Directing, and Controlling. Given the fact that most state workers' compensation programs are administrative in nature, appropriate planning can include, but is not limited to, such activities as the acquisition of the appropriate forms, development of status tracking mechanisms, establish communication lines with the local medical community, and informing employees of their rights and responsibilities under the workers' compensation act. Organizing an effective workers' compensation system can include, but is not limited to: selection and training of personnel who will be responsible for completing the appropriate forms, coordination with insurance or self-insured administrators, acquisition of appropriate rehabilitation and evaluation services, and development of medical response mechanisms. The directing phase can include, but is not limited to, implementation of tracking mechanisms, on-site visitation by medical and legal communities, development of work-hardening programs, and installation of return-to-work programs. Controlling can include such activities as the establishment of an audit mechanism to evaluate case status and progress of the program, use of injured worker home visitation, acquisition of outside investigation services, among other activities.
3. Compliance with the workers' compensation rules and regulations must be of the highest priority at all times. Appropriate training and education of individuals working within the workers' compensation management system should be mandatory and appropriate supervision should be provided at all times.
4. First things must be first. When an employee incurs a work-related injury or illness, appropriate medical treatment should be quickly provided. In some states, the employee possesses the first choice of a physician while in other states the employer has this choice. The injured or ill employee should be provided the best possible care in the appropriate medical specialty or medical facility as soon as feasible. Improper care in the beginning can lead to a longer healing period and additional costs.
5. Employers often fool themselves by thinking that if employees

are not told their rights under the state workers' compensation laws there is less chance that an employee will file a claim. This is a falsehood. In most states, employees have easy access to information regarding their rights under workers' compensation through the state workers' compensation agency, through their labor organization, or even through television commercials. A proactive approach that has proved successful is for the safety and resource control professional or other representative of the employer to explain to employees their rights and responsibilities under the workers' compensation laws of the state as soon as feasible after an injury, is sustained. This method alleviates much of the doubt in the minds of injured employees, begins or continues the bonds of trust, eliminates the need for outside parties being involved, and tends to improve the healing process.

6. The safety and resource control professional should maintain an open line of communication with the injured employee and attending physician. The open line of communications with the injured employee should be of a caring and informative nature and should never be used for coercion or harassment purposes. The open line of communications with the attending physician can provide the vital information regarding the status of the injured employee and any assistance the employer can provide to expedite the healing process.
7. Timely and accurate documentation of the injury or illness and appropriate filing of the forms to ensure payment of benefits is essential. Failure to provide the benefits in a timely manner as required under the state workers' compensation laws can prompt the injured employee to seek outside legal assistance and cause a disruption in the healing process.
8. Appropriate, timely, and accurate information should be provided to the insurance carrier, organization team members, and others to ensure that the internal organization is fully knowledgeable regarding the claim. There is nothing worse than an injured employee receiving a notice of termination from personnel while lying in a hospital bed, as can happen when personnel is not informed of the work-related injury and counts the employee absent from work.

9. As soon as medically feasible, the attending physician, insurance administrator, the injured employee, and the safety and resource control professional can discuss a return to light or to restricted work. A prudent safety and resource control professional may wish to use photographs or videotape of the particular restricted duty job, written job descriptions, and other techniques in order to ensure that all parties have a complete understanding of the restricted job duties and requirements. Once the injured employee has returned to restricted duty, the safety and resource control professional should see that the employee performs only the duties agreed upon, and within the medical limitations proscribed by the attending physician. A effective return-to-work program can be one of the most effective tools in minimizing the largest cost factor with most injuries or illnesses, namely time-loss benefits.
10. In coordination with the injured employee and attending physician, a rehabilitation program or work hardening program can be used to assist the injured employee to return to active work as soon as medically feasible. Rehabilitation or work-hardening programs can be used in conjunction with a return-to-work program.
11. Where applicable, appropriate investigative methods and services can be used to gather the necessary evidence to address fraudulent claims, deny non-work related claims, or to address malingering or other situations.
12. A prudent safety and resource control professional should audit and evaluate the effectiveness of the workers' compensation management program on a periodic basis to ensure effectiveness. All injured or ill employees should be appropriately accounted for, the status of each meticulously monitored, and cost factors continuously evaluated. Appropriate adjustments should be made to correct all deficiencies and to ensure continuous improvement in the workers' compensation management system.

APPENDIX H

OSHA REGIONAL OFFICES—DIRECTORY

Region I: Connecticut, Maine, Massachusetts, New Hampshire, Rhode Island, Vermont

1 Dock Square Building, 4th Floor
16–18 North Street
Boston, Massachusetts 02109
Phone: (617) 223-6710

Region II: New Jersey, New York, Puerto Rico, Virgin Islands

1515 Broadway, Room 3445
New York, New York 10036
Phone: (212) 944-3437

Region III: Delaware, District of Columbia, Maryland, Pennsylvania, Virginia, West Virginia

Gateway Building, Suite 2100
3535 Market Street
Philadelphia, Pennsylvania 19104
Phone: (215) 596-1201

Region IV: Alabama, Florida, Georgia, Kentucky, Mississippi, North Carolina, South Carolina, Tennessee

1375 Peachtree St. N.E., Suite 587
Atlanta, Georgia 30367
Phone: (404) 347-3573

Region V: Illinois, Indiana, Michigan, Minnesota, Ohio, Wisconsin
230 South Dearborn Street, Rm. 3244
Chicago, Illinois 60604
Phone: (312) 353-2220

Region VI: Arkansas, Louisiana, New Mexico, Oklahoma, Texas
525 Griffin Square Bldg., Rm. 602
Dallas, Texas 75202
Phone: (214) 767-4731

Region VII: Iowa, Kansas, Missouri, Nebraska
911 Walnut Street, Room 406
Kansas City, Missouri 64106
Phone: (816) 374-5861

Region VIII: Colorado, Montana, North Dakota, South Dakota, Utah, Wyoming
Federal Building, Room 1554
1961 Stout Street
Denver, Colorado 80294
Phone: (303) 837-3061

Region IX: Arizona, California, Hawaii, Nevada, American Samoa, Guam, Trust Territory of the Pacific Islands
450 Golden Gate Ave., Box 36017
San Francisco, California 94102
Phone: (415) 556-7260

Region X: Alaska, Idaho, Oregon, Washington
Federal Office Building, Room 6003
909 First Avenue
Seattle, Washington 98174
Phone: (206) 442-5930

APPENDIX I

POTENTIAL SOURCES OF SAFETY AND LOSS PREVENTION ASSISTANCE THROUGH LOCAL COLLEGES AND UNIVERSITIES

Auburn University

Leo A. "Tony" Smith, Professor

BS/MIE/MS/PCD, Industrial Engineering, Concentration in Safety and Ergonomics

Industrial Engineering Department
College of Engineering
Auburn University
207 Danston Hall
Auburn, AL 36849-5346
205-844-1415

The University of Alabama—Tuscaloosa

Dr. Paul S. Pay, Assistant Professor

BS/MS, Industrial Engineering

Industrial Engineering Department
The University of Alabama—Tuscaloosa
P.O. Box 870288
Tuscaloosa, AL 35487-0288
205-346-1603

University of Alabama—Birmingham

Joan Gennin, Program Administrator

MPH/PhD/DrPH, Environmental Health Sciences; MSPH/PhD, Industrial Hygiene; MPH/DrPH, Occupational Health and Safety; MSPH/PhD, Environmental Toxicology

Department of Environmental Health Sciences
School of Public Health
University of Alabama—Birmingham
Titmell Hall
Birmingham, AL 35294-0008
205-934-8488

Jacksonville State University

J. Fred Williams, Program Director

Occupational Safety and Health Technology

Department of Technology
Jacksonville State University
Room 217 Self Hall
700 Pelham Road North
Jacksonville, AL 36265
205-782-5080

University of North Alabama

Dr. Robert Gaunder, Professor

BS, Industrial Hygiene

Chemistry/Industrial Hygiene
University of North Alabama
UNA Box 5049
Florence, AL 35632
205-760-4474

Gate Way Community College

Ginger Jackson, Program Director

AA Occupational Safety and Health

Industrial Technology Division
Gateway Community College
108 N. 40th Street
Phoenix, AZ 85034
602-392-5000

Southern Arkansas University

James A. Collier, Program Head

BS, Industrial Technology

School of Science and Technology
Southern Arkansas University
100 East University
Magnolia, AR 71753
501-235-4284

University of California—Berkeley

Jeanne Bronk, Coordinator

MS/MPH/PhD, Environmental Health

Environmental Health Science Program
School of Public Health
University of California—Berkeley
Berkeley, CA 94720
510-643-5160

California State University—Fresno

Dr. Sanford Brown, Advisor

BS, Environmental Health

Environmental Health Science Program
California State University—Fresno
2345 E. San Ramon
Fresno, CA 93740-0030
209-278-4747

California State University—Fresno

Dr. Michael Waite, Advisor

BS, Occupational Safety and Health

Occupational Safety and Health Program
California State University—Fresno
2345 E. San Ramon
Fresno, CA 93740-0030
209-278-5093

University of Southern California

William J. Petak, Professor

BS/MS, Safety and Health

Institute of Safety and Systems Management Building
University of Southern California
University Park
Los Angeles, CA 90089-0021
213-740-2411

California State University—Los Angeles

Dr. Carlton Blanton, Professor

BS, Health Science; BS, Occupational Safety and Health; MA, Occupational Safety and Health; Certificate, Occupational Safety and Healts; Certificate, Environmental Health Certificate, Alcohol and Drug Problems

Health and Science Department
California State University—Los Angeles
5151 State University Drive
Los Angeles, CA 90032
213-343-4740

California State University—Northridge

Brian Malec, Chair

BS/MS, Environmental Health

Health Science Department
California State University—Northridge
18111 Nordhoft Street
Northridge, CA 91330
818-885-3100

Merritt College

Larry Gurley, Assistant Dean

AS, Occupational Safety and Health

Technical Division
Merritt College
12500 Campus Drive
Oakland, CA 94619
510-436-2409

National University

Ernest Wendi, Program Chair

BS/MS, Occupational Health and Safety Management and Technology

Department of Computers and Technology
Suite 205
National University
4141 Camino Del Rio South
San Diego, CA 92108
619-563-7124

Colorado State University

Kenneth Blehm, Coordinator

BS/MS/PhD, Environmental Health

Department of Environmental Health
College of Veterinary Medicine and Biomedicine
Colorado State University
Fort Collins, CO 80523
303-491-7038

Red Rocks Community College

Anne—Mario Edwards, Department Coordinator

MS, Occupational Safety Technology Certificate, Occupational Safety Technology

Department of Occupational Safety Technology
Red Rocks Community College

Campus Bon 41
13300 W. 6th Avenue
Lakewood, CO 80401-5398
303-914-6338

Trinidad State Junior College

Charles McGlothlin, Associate Professor

AAS, Occupational Safety and Health Certificate, Occupational Safety and Health

Occupational Safety Department
Trinidad State Junior College
600 Prospect Street
Trinidad, CO 81082
719-846-5502

Central Connecticut State University

Andrew Baron, Assistant Dean

BS, Occupational Safety and Health; BS, Public Safety

Occupational Safety Health Department
School of Technology
Central Connecticut State University
1615 Stanley Street
New Britain, CT 06050
203-827-7997

University of New Haven

Garher, Director

AS/BS, Occupational Safety and Health Administration; AS/BS, Occupational Safety and Health Technology; MS, Occupational Safety and Health Management; MS, Industrial Hygiene

Department of Occupational Safety and Health
University of New Haven
300 Orange Avenue
West Haven, CT 06516
203-932-7175

Florida International University

Gabriel Aurioles, Professor

BS/MS, Construction Management

Construction Management Department
Florida International University
University Park VH230
107th and 8th Avenue
Miami, FL 33199
305-348-3542

Miami-Dade Community College

Wilfred J. Muniz, Director

AS, Fire Science Technology AS, Fire Science Administration

Fire Science Technology
Academy of Science
Miami—Dade Community College
11380 NW 27th Avenue
Miami, FL 33167
305-237-1400

Hillsborough Community College

Keith Day, Coordinator

AS, Fire Science Technology

Fire Safety Department
Hillsborough Community College
RD Box 5096
Tampa, FL 33675-5096
813-253-7628

University of Florida

Richard Coble, PhD, Associate Professor

MS, Building Construction, Concentration in Construction Safety

M.E. Rinker Senior School of Building Construction
University of Florida

FAC 100/BON
Gainesville, FL 32611-2032
352-392-7521

University of Florida

Dr. Joseph J. Delsino, Chair
BS/MS/PhD, Environmental Engineering
Department of Environmental Engineering Sciences
University of Florida
P.O. Box 116450
Gainesville, FL 32611-6450
352-392-0841

University of Georgia

Harold Barnhart, Coordinator
BS, Environmental Health Science
Environmental Health Science
University of Georgia
Room 206, Dairy Science Building
Athens, GA 30602-2102
706-542-2454

Georgia Institute of Technology

Dr. Leland Riggs, Associate Director/Academic
MS/PhD, Environmental Engineering
Graduate Program of Environmental Engineering
School of Civil Engineering
Georgia Institute of Technology
790 Atlantic Drive
Atlanta, GA 30332
404-894-2000

University of Hawaii

Arthor Kodama, Department Chair
MS/MPH, Environmental Health
Environmental and Occupational Health Program

Department of Public Health Sciences
School of Public Health
University of Hawaii
1960 East-West Road
Honolulu, HI 96822
808-956-7425

Southern Illinois University—Carbondale

Keith Contor, Associate Professor

BS, Industrial Technology

Department of Technology
Southern Illinois University
Carbondale, IL 62901
618-536-3396

University of Illinois—Chicago

Dr. William Hallenbeck, Director

MS/PhD, Safety Engineering; MS/PhD, Environmental Health; MS/PhD, Industrial Hygiene; MS/PhD, Industrial Safety

Industrial Hygiene Programs Environmental and Occupational Health Sciences
School of Public Health West
University of Illinois—Chicago
2121 W. Taylor
Chicago, IL 60612
312-996-8855

Northern Illinois University

Earl Hansen, Chair

BS, Industrial Technology, Concentration in Safety MS, Industrial; Management, Concentration in Safety or Industrial Hygiene; PhD, Education, Concentration in Safety

Department of Technology
Northern Illinois University
Still Hall, Room 203
DeKalb, IL 60115-1349
815-753-0579

Western Illinois University

Dan Sigwart, Professor

BS, Health Science Minor in Industrial Safety

Health Sciences Department
Western Illinois University
402 Stipes Hall
Macomb, IL 61455
309-298-2240

Illinois State University

Edmond Corner, Director

BS, Safety; BS, Environmental Health

Safety Studies
Department of Health Sciences
College of Applied Science and Technology
Illinois State University
Mail Code 5220
Normal, IL 61790-5220
309-438-8329

University of Illinois—Champaign

Vernon Snoeyink, Supervisor

BS/MS, Civil Engineering, Environmental Emphasis

Environmental Engineering and Science Program
Civil Engineering Department
3230 Newmark CE Laboratory
University of Illinois—Champaign
205 N. Matthews
Urbana, IL 61801
217-333-6968

Indiana University

James W. Crowe, Chair

AS, Hazard Control; BS, Occupational Safety and Health; MS, Safety Management; HSD, Safety Education

Hazard Control Program

Applied Health Science/HCP
School of Health, Physical Education and Recreation
Indiana University—Bloomington
HPER 116
Bloomington, IN 47405
812-855-2429

Indiana State University

John Doty, Chair

BS, Safety Management; BS, Environmental Health; MS, Health and Safety

Industrial Health and Safety Management Program
Applied Health Science Department
School of Health, Physical Education and Recreation
Indiana State University
Terre Haute, IN 47809
812-237-3079

Purdue University

Dr. Paul Ziemer, Department Head

BS, Environmental Health; BS, Environmental Engineering; BS/MS/PhD, Industrial Hygiene; BS/MS/PhD, Health Physics

School of Health Sciences
Purdue University
1163 Civil Engineering Building
West Lafayette, IN 47907
317-494-1392

Purdue University

William E. Field, Professor

MS/PhD, Agricultural Safety and Health

Department of Agricultural Engineering
Purdue University
1146 Agricultural Engineering Building
West Lafayette, IN 47907-1146
317-494-1173

Iowa State University

Jack Beno, Coordinator

BS, Occupational Safety and Health

Occupational Safety Program
School of Education
Iowa State University
Industrial Education Building 2
Room 122
Ames, IA 50010
515-294-5945

Western Kentucky University

Donald Carter, Coordinator

AS, Occupational Safety and Health; BS, Industrial Technology; Concentration in Occupational Safety and Health

Occupational Health and Safety Program
Department of Public Health
Western Kentucky University
1 Big Red Way
Bowling Green, KY 42101
502-745-5854

Morehead State University

Dr. Brian Reeder, Coordinator

BS, Environmental Studies

Department Biological–Environmental Sciences
Morehead State University
Morehead, KY 40351
606-783-2945

Murray State University

David G. Kraemer, Chair

BS/MS, Occupational Safety and Health

Occupational Safety and Health Department

Murray State University
P.O. Box 9
Murray, KY 42071
502-762-2488

Eastern Kentucky University

Larry Collins, Coordinator

AA, Fire and Safety; BS, Fire and Arson; BS, Industrial Risk Management; BS, Fire Protection Administration; BS, Fire Protection Engineering Technology; BS, Insurance and Risk Management; MS, Loss Prevention and Safety

Fire and Safety Engineering Technology Program
Loss Prevention and Safety Department
College of Lam Enforcement
Eastern Kentucky University
220 Stratton Building
Richmond, KY 40475
606-622-1051

Louisiana State University

Lalit Verma, Department Head

BS, Industrial and Agricultural Technology

Department of Agriculture—Engineering
Louisiana State University
Baton Rouge, LA 70803
504-388-3153

Nicholls State University

Michael Flowers, Program Coordinator

AS, Petroleum Safety

Petroleum Services Department
Nicholls State University
P.O. Box 2148
University Station
Thibodaux, LA 70301
504-448-4740

University of Southwestern Louisiana

Thomas E. Landry, Associate Professor

BS, Industrial Technology, Concentration in Safety

Department of Industrial Technology
University of Southwestern Louisiana
P.O. Box 42972
Lafayette, LA 70504
318-482-6968

Central Maine Technical College

Patricia Vampatella, Assistant Dean

AAS, Applied Science

Occupational Health and Safety Department
Central Maine Technical College
1250 Turner Street
Auburn, ME 04210
207-784-2385

Johns Hopkins University

Dr. Patrick Breysse, Director

MHS/PhD, Environmental Health, Engineering and Safety Sciences; MHS/PhD, Industrial Hygiene and Safety Sciences

School of Hygiene and Public Health Environmental Sciences
Johns Hopkins University
615 N. Wolfe Street
Baltimore, MD 21205
410-955-3602

Salisbury State University

Elichia A. Venso, PhD, Assistant Professor

BS, Environmental Health

Environmental Health Department
Salisbury State University
Salisbury, MD 21801
410-543-6490

University of Maryland

Dr. Steven Spivak, Chair

BS, Fire Protection Engineering; MS, Fire Protection Engineering; ME, Fire Protection Engineering

Department of Fire Protection Engineering Room 0151, Engineering Classroom Building
A. James Clark School of Engineering
Glenn L. Martin Institute of Technology
University of Maryland
College Park, MD 20742-3031
301-405-6651

North Shore Community College

Frank Ryan, Chair

AA, Fire Protection Safety Technology

Fire Protection Safety Department
North Shore Community College
1 Ferncroft Road
Danvers, MA 01923
508-762-4000, ext. 5562

University of Massachusetts

Dr. Michael Ellenbecker, Coordinator

MS, Engineering, Concentration in Industrial Hygiene and Ergonomics; MS/ScD, Engineering, Concentration in Work Environments and Safety Ergonomics

Work Environments Department
University of Massachusetts
1 University Avenue
Lowell, MA 01854
508-934-3250

Tufts University

John Kreilfeldt, Professor

BS, Engineering Psychology; MS/PhD, Human Factors

Human Factors Program
Mechanical Engineering Department
College of Engineering
Tuffs University
Anderson Hall
Medford, MA
617-628-5000, ext. 2209

Worcester Polytechnic Institute

David Lucht, Director

MS/PhD, Fire Protection Engineering

Fire Protection Engineering Center for Fire Safety Studies
Worcester Polytechnic Institute
100 Institute Road
Worcester, MA 01609
508-831-5593

Henry Ford Community College

Sally Goodwin, Director

AS, Fire Science; AA, Property Assessment

Management Development Division
Henry Ford Community College
22586 Ann Arbor Trail
Dearborn Heights, M1 48127
313-730-5960

Wayne State University

Dr. David Bassett, Chair

MS, Occupational and Environmental Health

Occupational and Environmental Health Sciences
College of Pharmacy and Allied Health
Wayne State University
628 Shapero Hall
Detroit, M1 48202
313-577-1551

Madonna University

Florence Schaldenbrand, Chair

AS/BS, Occupational Safety, Health and Fire Science

Physical and Applied Sciences
College of Science and Mathematics
Madonna University
36600 Schoolcraft Road
Livonia, MI 48150-1173
313-591-5110

Central Michigan University

Louis Ecker, Professor

BS, Applied Science; Minor in Industrial Safety; MS, Industrial Management and Technology

Department of Industrial and Engineering Technology
Central Michigan University
Mount Pleasant, MI 48859
517-774-6443

Oakland University

Dr. Sherryl Schutz, Director

BS, Industrial Safety

Industrial Health Program
School of Health Sciences
Oakland University
Rochester, MI 48309-4401
313-370-4038

Grand Valley State University

Dr. Eric Van Fleet, Director

BS, Occupational Safety and Health

Occupational Safety and Health Program
School of Health Sciences
Grand Valley State University
1 Campus Drive

Allendale, M1 49401-9403
616-895-3318

University of Michigan—Ann Arbor

Frances Bourdas, Graduate Program Assistant
BS/MS/MSE/PhD, Industrial and Operations Engineering; MS, Engineering/Occupational Ergonomics
Industrial Operations Engineering
University of Michigan—Ann Arbor
1205 Beal Avenue, IDE Building
Ann Arbor, M1 48109-2117
313-764-6480

University of Michigan—Ann Arbor

Dr. Richard Garrison, Director
MS/MPH, Industrial Hygiene; MPH/PhD, Environmental Health
Environmental and Industrial Health Department
School of Public Health
University of Michigan—Ann Arbor
Ann Arbor, M1 48109
313-764-2594

Ferris State University

Lori A. Seller, Assistant Professor
BS, Industrial Safety and Environmental Health
College of Applied Health Sciences
Ferris State University
200 Ferris Drive
Big Rapids, M1 49307
616-592-2307

University of Minnesota—Duluth

Bernard DeRobels, Director
MIS, Industrial Hygiene; MIS, Industrial Safety
Master of Industrial Safety Program
Department of Industrial Engineering

University of Minnesota—Duluth
105 Voss-Kovach Hall
Duluth, MN 55812
218-726-8117

University of Minnesota

Kathy Soupir, Coordinator

MS/PhD, Environmental Health

Environmental and Occupational Health
University of Minnesota
School of Public Health
RD Box 807, UMHC
Minneapolis, MN 55455
612-625-0622

University of Southern Mississippi

Dr. Emmanuel Ahua, Program Director

MPH, Public Health, Concentration in Occupational and Environmental Health

Center for Community Health
College of Health and Human Sciences
University of Southern Mississippi
Box 5122 Southern Station
Hattiesburg, MS 39406-5122
601-266-5437

Central Missouri State

Dr. John J. Prince, Department Head

BS, Safety Management; BS/MS, Industrial Hygiene; MS, Transportation Safety; MS, Fire Science; MS, Public Service Administration; MS, Security; MS, Industrial Safety Management; ED, Safety

Safety Science and Technology Department
Central Missouri State
Humpreys Building, Room 325
Warrensburg, MO 64093
816-543-4626

St. Louis Community College—Forest Park

Emil Hrbacek, Coordinator

AA, Fire Protection Safety

Municipal Services
St. Louis Community College—Forest Park
5600 Oakland
St. Louis, MO 63110
314-644-9310

Montana Tech

Julie B. Norman, CIH, Department Head

AS, Occupational Safety and Health; BS, Occupational Safety and Health; BS, Environmental Engineering; MS, Industrial Hygiene

Occupational Safety and Health/Industrial Hygiene Department Montana Tech
1300 W. Park Street
Butte, MT 59701
406-496-4393

University of Nebraska—Kearney

Darrel Jensen, Director

BS, Safety Education; BS, Occupational Safety and Health; BS, Transportation Safety; BS, Driver Education

Nebraska Safety Center
University of Nebraska—Kearney
West Center
Kearney, NE 68849
308-234-8256

Community College of Southern Nevada

Sonny Lyerly, Chair

AAS, Fire Science Technology

Department Mathematics, Health and Human Services
Community College of Southern Nevada
6375 W. Charleston

Las Vegas, NV 89102
702-643-6060, ext. 439

Keene State College

David Buck, Director

AS, Chemical Dependency; BS, Industrial Safety; BS, Occupational Safety and Health

Safety Center
Keene State College
229 Main Street
Keene, NH 03431
603-358-2977

Camden County College

Matthew Davies, Coordinator

AA, Occupational Safety; AA, Fire Science

Information Services
Camden County College
RD Boa 200
Blackwood, NJ 08012
609-227-7200, ext. 251

Rutgers, The State University of New Jersey

Frank Haughey, Director

BS/MS, Radiation Science

Radiation Science Program, Rutgers, The State University of New Jersey
Building 4087, Livingston Campus
New Brunswick, NJ 08093
908-932-2551

New Jersey Institute of Technology

Howard Gage, Director/Associate Professor

MS, Occupational Safety and Health

Occupational Safety and Health Department of Mechanical and Industrial Engineering
New Jersey Institute of Technology

University Heights
Newark, NJ 07102
201-596-3653

Thomas Edison State College

Janice Touver, Admissions

AS/BS, Fire Protection Science; AS/BS, Environmental Science and Technology; AS/BS, Industrial Engineering Technology

Applied Science and Technology
Thomas Edison State College
101 W. State Street
Trenton, NJ 08608-1176
609-984-1150

New Mexico Institute of Mining and Technology

Dr. Clint Richardson, Associate Professor

BS, Environmental Engineering

Department of Mineral and Environmental Engineering
New Mexico Institute of Mining and Technology
Campus Station
801 Leroy
Socorro, NM 87801
505-835-5345

Broome Community College

Francis Short, Chair

AAS, Fire Protection Safety

Special Career Programs Department
Broome Community College
RD Box 1017
Binghamton, NY 13902
607-778-5000

Mercy College

Dr. Joe Sullivan, Chair

BS, Public Safety Certificates: Fire Science, OSHA, Public Safety, Private Security

Criminal Justice and Public Safety Department
Mercy College
Social Science Building
555 Broadway
Dobbs Ferry, NY 10522
914-674-7320

New York University

Katie B. Shadow, Graduate Coordinator

MS, Occupational and Industrial Hygiene; PhD, Environmental Health Sciences

Environmental Health Sciences Program
Nelson Institute of Environmental Medicine
New York University
A.J. Lanza Laboratories
Long Meadow Road
Tuxedo, NY 10987
914-351-5480

State University of New York College of Technology

John Tiedemann, Department Head

BS, Industrial Technology

Department of Industrial Technology—Facility Management
Technology State University of New York College of Technology
Route 110
Farmingdale, NY 11735
516-420-2326

Columbia University

Anne Hutzelmann, Administrative Assistant

MS/DrPH, Public Health

Division of Environmental Sciences
Columbia University
188th Street
New York, NY 10032
212-305-3464

University of Rochester

Mary Wahlman, Coordinator

MS, Environmental Studies; MS, Industrial Hygiene

Department of Biophysics
School of Medicine
University of Rochester
Rochester, NY 14642
716-275-3891

University of North Carolina—Chapel Hill

David Leith, Program Director

BSPH/PhD, Environmental Science and Policy; MS, Public Health

Environmental Sciences and Engineering
School of Public Health
University of North Carolina—Chapel Hill
201 Columbia Street
Chapel Hill, NC 27599-7400
919-966-3844

University of North Carolina—Chapel Hill

Larry Hyde, Deputy Director

Continuing Education in Occupational Health and Safety Private Seminars

Research Center
University of North Carolina—Chapel Hill
109 Conners Drive, #1101
Chapel Hill, NC 27514
919-962-2101

Central Piedmont Community College

Andy Nichols, Director

AA, Industrial Safety

Industrial Safety
Central Piedmont Community College
RD Box 35009
Charlotte, NC 28235-5009
704-342-6582

Western Carolina University

Robert Dailey, Coordinator

BS, Electronics Engineering Technology; BS, Industrial Technology; BS, Manufacturing Engineering Technology; BS, Industrial Distribution MS, Technology

Occupational Safety Program
Industrial and Engineering Technology Department
Western Carolina University
226 Belk Building
Cullowhee, NC 28723
704-227-7272

North Carolina A&T State University

Dillip Shah, Coordinator

BS, Occupational Safety and Health

Department of Construction Management and Safety
North Carolina A&T State University
Price Hall, Room 124
Greensboro, NC 27411
919-334-7586

East Carolina University

Dr. Mark Friend, Program Director

BS, Environmental Health, Option in Industrial Hygiene MSIT Occupational Safety

Department of Industrial Technology
East Carolina University
105 Flanagan
Greenville, NC 27858
919-328-4249

North Carolina State University

Richard G. Pearson, Professor

PhD, Industrial Engineering; Concentration in Ergonomics

Department of Industrial Engineering
North Carolina State University
Box 7906

Raleigh, NC 27695
919-515-6410

North Dakota State College of Science

Linda Johnson, Instructor

AS, Industrial Hygiene; AS, Occupational Health and Safety

North Dakota State College of Science
800 N. 8th Street
Wahpeton, ND 58076
701-671-2202

University of Akron

Dr. David H. Hoover, Program Head

AAS, Fire Protection Technology, 2+2 Option in Technical Education; BS, Fire Protection

Fire Protection Program
Division of Public Service Technology
The University of Akron
Akron, OH 44325-4304
216-972-7789

University of Cincinnati

William M. Kraemer, Director

AAS, Fire Science Technology; BS, Fire Science Engineering

College of Applied Science
University of Cincinnati
2220 Victory Parkway, ML 103
Cincinnati, OH 45206
513-556-6583

University of Cincinnati

Dr. Rod Simmons, Assistant Research Professor

MS/PhD, Industrial Engineering, Concentration in Occupational Safety

Department of Mechanical, Industrial and Nuclear Engineering
University of Cincinnati
Mail Location 116

Cincinnati, OH 45221-0116
513-556-2738

Stark Technical College

Cameron H. Speck, Program Developer

AS, Engineering Technology; AS, Allied Health

Safety/Risk Management
Continuing Education
Stark Technical College
6200 Frank Avenue, N.W.
Canton, OH 44720
216-494-6170

Wright State University

Allan Burton, Director

BS, Environmental Sciences

Environmental Health Sciences Program
Biological Sciences Department
Wright State University
Colonel Glenn Highway
Dayton, OH 45435
513-873-2655

Wright State University

Dr. Jennie Gallimore, Associate Professor

BS, Human Factors Engineering

Department of Biomedical and Human Factors Engineering
College of Engineering
Wright State University
207 Russ Center
Dayton, OH 45435
513-873-5044

East Central University

Dr. Paul Woodson, Chair

BS, Environmental Science, Concentrations in Environmental Health, Industrial Hygiene and Environmental Management

Environmental Science Program
Physical and Environmental Sciences
Department
East Central University
Ada, OK 74820
405-332-8000, ext. 547

University of Central Oklahoma

Dr. Lou Ebrite, Department Chair

BS, Industrial Safety

Occupational and Technology Education Department
College of Education
University of Central Oklahoma
100 N. University Drive
Edmond, OK 73034-0185
405-341-5009

University of Oklahoma

Deborah Imel Nelson, Program Head

MS, Environmental Science

Civil Engineering and Environmental Science Department
University of Oklahoma
202 W. Boyd Street, Room 334
Norman, OK 73109
405-325-5911

University of Oklahoma—Oklahoma City

Robert Nelson, Associate Professor

MS/MPH, Environmental Management; MS/MPH, Environmental Toxicology; MS/MPH, Industrial Hygiene

Occupational and Environmental Health Department
University of Oklahoma—Oklahoma City
P.O. Box 26901
Oklahoma City, OK 73190
405-271-2070

Southeastern Oklahoma State University

Robert Semonisck, PhD, Professor, Safety

BS, Occupational Safety and Health

School of Applied Science and Technology
Southeastern Oklahoma State University
Station A
Durant, OK 74702
405-924-0121, ext. 2464

Oklahoma State University

Dr. Don Adams, Coordinator

BS, Fire Protection and Safety Engineering Technology

Fire Protection and Safety Technology Department
303 Campus Fire Station
Oklahoma State University
Stillwater, OK 74078
405-744-5639

Mount Hood Community College

Dr. David Mohtasham, Coordinator

AAS, Environmental Safety and Hazardous Materials Management

ESHM Program
Mount Hood Community College
Route 26,000 S.E. Stark Street
Gresham, OR 97030
503-667-7440

Southwestern Oregon Community College

Darryl Saxton, Coordinator

AAS, Fire Protection

Fire Science Program
Southwestern Oregon Community College
Coos Bay, OR 97420
503-888-2525

Oregon State University

Dave Lawson, Associate Professor

BS, Environmental Health and Safety; MS, Safety Management; MS, Environmental Health Management, Concentration in Occupational Safety; PhD, Health

Safety Studies Program
Department of Public Health
College of Health and Human Performance
Oregon State University
Waldo Hall, Room 256
Corvallis, OR 97331-6406
503-737-2686

Indiana University of Pennsylvania

Dr. Robert Soule, Chair

BS/MS, Safety Sciences

Safety Science Department
College of Health and Human Science
Indiana University of Pennsylvania
117 Johnson Hall
Indiana, PA 15705
412-357-3019

Millersville University of Pennsylvania

Dr. Paul Specht, Coordinator

BS, Occupational Safety and Hygiene Management

Department of Industry and Technology
Millersille University of Pennsylvania
P.O. Box 1002
Millersille, PA 17551
717-872-3981

Northampton Community College

Kent Zimmerman, Program Director

AAS, Applied Science, Concentrations in Safety, Health and Environmental Technology

Safety, Health and Environmental Technology
Northampton Community College
3835 Green Pond Road
Bethlehem, PA 18017
610-861-5590

Slippery Rock University of Pennsylvania

Dr. Joseph Calli, Chair

BS, Safety and Environmental Management
Allied Health Department
Slippery Rock University of Pennsylvania
Behavioral Science Building, Room 208
Slippery Rock, PA 16057
412-738-2017

Francis Marion University

Dr. W.H. Breazeale, Department Head

BS, Health Physics
Department of Chemistry and Physics
Francis Marion University
RD Box 100547
Florence, SC 29501
803-661-1440

University of South Carolina—Columbia

Dr. Edward Oswald, Professor

MSPH/MPH/PhD, Occupational Health, Environmental Duality and Hazardous Materials Management
Department of Environmental Health
School of Public Health Sciences
University of South Carolina—Columbia
Health Sciences Building, Room 311B
Columbia, SC 29208
803-777-4120

East Tennessee State University

Creg Bishop, Interim Chair

BS/MS, Environmental Health Minor in Safety

Environmental Health Department
College of Public and Allied Health
East Tennessee State University
Johnson City, TN 37614
615-929-4268

Middle Tennessee State University

Dr. Richard Redditt, Professor

MS, Industrial Studies, Concentration in Safety

Industrial Studies Department
Middle Tennessee State University
P.O. Box 19
Murfreesboro, TN 37132
615-898-2776

University of Tennessee—Knoxville

Charles Hamilton, Chair

BS/MS/EdD/PhD, Health Education; MS/EdS, Safety Education; MS, Public Health

Health, Leisure and Safety Department
University of Tennessee—Knoxville
1914 Andy Holt Drive
Knoxville, TN 37996-2700
615-974-6041

Lamar University

Dr. Victor Zalcom, Department Chair

BS, Industrial Technology; BS, Industrial Engineering

Industrial Engineering
Lamar University
P.O. Box 10032-LUS
Beaumont, TX 77710
409-880-8804

University of Houston—Clearlake

Dr. Dennis Casserly, Associate Professor

BS, Environmental Science

Division of Natural Sciences
University of Houston—Clearlake
2700 Bay Area Boulevard
Houston, TX 77058
713-283-3775

Texas A & M University

Dr. James Rock, Associate Professor

BS/MS, Safety Engineering; BS/MS, Industrial Hygiene; BS/MS, Health Physics

Safety Division
Nuclear Engineering
Texas A & M University
College Station, TX 77843-3133
409-862-4409

Texas Tech University

Dr. Mica Endsley, Assistant Professor

BS/MS/PhD, Industrial Engineering, Concentration in Ergonomics

Department of Industrial Engineering
Texas Tech University
P.O. Box 43061
Lubbock, TX 79409
806-742-3543

Sam Houston State University

Dr. James R. DeShaw, Program Head

BS, Environmental Sciences

Department of Biological Sciences
Sam Houston State University
P.O. Box 2116
Huntsville, TX 77341-2116
409-294-1020

San Jacinto College Central

Gary M. Vincent, Chair

AA, Occupational Health and Safety Technology

Division of Health Science
Health and Safety Technology Department
San Jacinto College Central
8060 Spencer Highway
Pasadena, TX 77501-2007
713-476-1834

The University of Texas at Tyler

Dr. W. Clayton Allen, Chair

BS/MS, Technology, Concentration in Industrial Safety

The University of Texas at Tyler
School of Education and Psychology
Department of Technology
3900 University Boulevard
Tyler, TX 74799
903-566-7331

Texas State Technical College

David Day, Department Chair

AAS, Occupational Safety and Health; AAS, Hazardous Materials Management; AAS, Radiation Protection Technician

Occupational Safety and Health Department
Texas State Technical College
3801 Campus Drive
Waco, TX 76705
817-867-4841

University of Utah

Donald S. Bloswick, Associate Professor

MS/ME/PhD, Mechanical Engineering, Concentration in Ergonomics and Safety; MPH/MSPH, Public Health, Concentration in Ergonomics and Safety

Mechanical Engineering Department

University of Utah
3209 MEB
Salt Lake City, UT 84112
801-581-4163
bloswick@me.mech.utah.edu

Virginia Commonwealth University

Michael McDonald, Coordinator

BS, Safety and Risk Control Administration

Safety and Risk Administration Program
Justice/Risk Administration Department
School of Community and Public Affairs
Virginia Commonwealth University
913 W. Franklin Street
Richmond, VA 23284
804-828-6237

Virginia Tech

Tom Dingus, Professor

MS, Safety Engineering

Department of Industrial Engineering
Virginia Tech University
302 Whittemore Hall
Blacksburg, VA 24061
540-231-8831

Central Washington University

Ronald Hales, Professor

BS, Loss Control Management; Minor in Traffic Safety; Minor in Loss Control Management

Industrial Engineering
Technology Department
Hebeler Hall
Central Washington University
Ellensburg, WA 98926
509-963-3218

University of Washington

Mary Lou Wager, Graduate Program Assistant
MS/PhD, Industrial Hygiene and Safety
Environmental Health Department School of Public Health and Community Medicine
Mail Stop SC-34
University of Washington
Seattle, WA 98195
206-543-3199

Fairmont State College

John Parks, Safety Coordinator
BS, Safety Engineering Technology
Technology Division
Fairmont State College
Locust Avenue
Fairmont, WV 26554
304-367-4633

West Virginia University

Terrence Stobbe, Director
MS, Occupational Hygiene and Occupational Safety
Department of Industrial Engineering
College of Engineering
West Virginia University
RD Box 6101
Morgantown, WV 26506-6101
304-293-4607

West Virginia University

Daniel E. Della-Guistina, Chair
MS, Safety and Environmental Management
Department of Safety and Environmental Management
West Virginia University
P.O. Box 6070, COMER
Morgantown, WV 26506
304-293-2742

Marshall University

Keith Barenklau, Program Director

BS/MS, Safety Technology, Occupational Safety Option; MS, Safety Technology, Safety Management Option; MS, Mine Safety

Safety Technology Department
Gullickson Hall, Room 3
College of Education
Marshall University
Huntington, WV 25755-2460
304-696-4664

University of Wisconsin—Eau Claire

Dale Taylor, Chair

BS, Environmental and Public Health; MS, Environmental and Public Health

Department of Allied Health Professions
University of Wisconsin—Eau Claire
Eau Claire, WI 54702-4004
715-836-2628

University of Wisconsin—Stout

John Olson, Director

MS, Occupational Safety and Health

Safety and Loss Control Center
Industrial Management Department
University of Wisconsin—Stout
205 Communications Center
Menomonie, WI 54751
714-232-2604

University of Wisconsin—Platteville

Roger Hauser, Professor

BS/MS, Industrial Technology, Management and Occupational Safety, Concentration in Safety

Industrial Studies Department
University of Wisconsin—Platteville
309 Pioneer Tower

Platteville, WI 53818
608-342-1187

University of Wisconsin—Steven's Point

Dr. Ann Abbott, Director

BS, Health Promotion and Safety Health Protection Minor in Safety
School of HPERA
University of Wisconsin—Steven's Point
131 Quandt
Steven's Point, WI 54481
715-346-4420

University of Wisconsin—Whitewater

Jerome W. Witherill, Chair

BS/MS, Safety Majors in Institutional Safety, Occupational Safety, and Traffic Safety[1]
Department Safety Studies
University of Wisconsin—Whitewater
800 W. Main Street
Whitemater, WI 53190
414-472-1117

ENDNOTE

1. American Society of Safety Engineers 1996–97 Survey of College and University Safety and Related Degree Programs.

INDEX